P9-BZH-862

Design of Mission Operations Systems for Scientific Remote Sensing

To the Magellan Flight Team, whose encouragement and support have enabled the book to be achieved, and whose discussions have allowed it to reflect the realities of what a mission is really like; and to Kate, who provided the inspiration and patiently tolerated the effort (SW).

To the members of the Magellan Spacecraft Team, without whose support the experience to write this book might never have been achieved (KL).

To Eric,
thanks for all your attitude control support during the Magellan mission.
Ken Leadbetter

Design of Mission Operations Systems for Scientific Remote Sensing

Stephen D. Wall
Jet Propulsion Laboratory, California Institute of Technology
Kenneth W. Ledbetter
Martin Marietta Astronautics Group

Taylor & Francis
London, Washington DC
1991

UK Taylor & Francis Ltd, 4 John St, London WC1N 2ET

USA Taylor & Francis Inc., 1900 Frost Road, Suite 101, Bristol, PA 19007

British Library Cataloguing in Publication Data

Wall, Stephen D.
Design of mission operations systems for scientific remote sensing.
I. Title II. Ledbetter, Kenneth W.
621.26

ISBN 0–85066–860–3

Library of Congress Cataloging in Publication Data is available

Printed in Great Britain by Burgess Science Press, Basingstoke, on paper which has a specified pH value on final paper manufacture of not less than 7.5 and is therefore 'acid free'.

Contents

Foreword

This volume is the first definitive description of the various models of mission operations systems (MOS) available to the Project Manager, the Operations Manager, or to the student, to aid in the understanding and design of a reasonable cost mission operations system for future unmanned, remote-sensing missions, either Earth observing or planetary. As more and more such missions have flight operations lifetimes measured in years, even tens of years, the MOS becomes an increasingly more difficult, more costly, and more human resource limited process. No longer can we afford to design the MOS last. In fact, the MOS concept may well drive the spacecraft design, the science instrument design, and the ground system, including new facilities.

The authors have done an excellent job in defining the basic elements of an MOS, and in providing guidance for selection of optimal arrangements. Future MOS designs must consider performance and cost as equal parameters, while keeping mission success and flexibility as required objectives.

It is hoped that this volume will encourage young engineers to participate, and to seek out opportunities to join MOS design teams, and to become flight team members. The innovative ideas and a willingness to change the old ways of doing business are needed today in mission operations design as much, if not more so, than in spacecraft design.

James S. Martin
Viking Project Manager (ret.)
September 1990

Preface

This book was written with two objectives. First, in the past 20 years enough remote-sensing missions have been flown that the process by which mission operations systems (MOS) are designed has matured sufficiently to justify documentation. Second, mission operations system designs themselves have evolved, through application of lessons learned to subsequent missions. It seems probable that the number of such missions will grow in the future, so we believe that the cumulative experience of past operations should be analysed and recorded as an archive to provide guidance for these future missions. The purpose of this book, then, is to provide a description of the mission operations systems design process and the general principles of the designs themselves. It is intended to be useful to professionals designing an MOS and for students of remote sensing or other spacecraft missions who need to know how one is organized and run.

Many other objectives are vital to provide insight about how modern missions are run. Certainly there are very few individuals who understand all the details of any mission, but without general concepts of such fields as spacecraft and flight software design, science objectives, remote-sensing methods, and radio frequency communications (just to name a few), knowledge of mission operations systems is sadly incomplete. Perhaps, then, a few words are in order here about what is not our purpose. Spacecraft design, for instance, has progressed significantly in the past decade, yet we speak of it in only an introductory way. In particular, as the distances which spacecraft travel from Earth have grown, the communication times (the time for signals to travel the distance) have become longer than is required for satisfactory reaction to on-board anomalies. Thus spacecraft systems have had to develop autonomous on-board responses. The design of such hardware and software fault protection mechanisms has become a research field of its own. Although we briefly mention this field here, a more complete comprehension than can be obtained from what we have provided is helpful if not a necessity to all MOS personnel when things begin to go wrong. Similarly, whereas early missions carried little more than television cameras as payloads, the imaging systems of recent missions are much more complex and involve differences even in the basic nature of how images are taken. Non-imaging instruments have similarly evolved. Complex remote-sensing sciences need more sophisticated remote sensors, and, although we review the basics here, there is much more to be said. Literature references are given where possible.

Remote-sensing science provides the all-important link between the data acquired in missions and the bodies under investigation. Entire series of books have been dedicated to that purpose. We have provided some references, but this field is maturing so quickly that the reader must often find his own way. At the heart of most remote sensing are the more fundamental science disciplines which attempt to understand the bodies themselves. Basic knowledge of all these fields is valuable to mission operations personnel. Finally let us mention that an impressive contributor to the successes of missions has been improvements in radio communications. In particular, the National Aeronautics and Space Administration's Deep Space Network and Tracking and Data Relay Satellite System have been able to yield in many cases higher communications rates to platforms than are available on Earth. These systems provide both the ability to command and to return data using remarkably low spacecraft power. To accomplish this they need very large antennas, some more than 60m (200 feet) across, which have advanced mechanical systems to precisely point them at their quarry. They also employ state-of-the-art electronic techniques to acquire miniscule signals in the presence of strong noise from both local and extraterrestrial sources. Operation of missions requires as much understanding of their methods as can be had, yet we provide a less than basic overview here. It is unfortunate that in most cases experience will be the only teacher of many such disciplines.

An operations system is defined as the system necessary to perform, monitor and control an operation. It may consist of people, hardware, software, and/or documentation according to the complexity of the operation. Some are highly repetitive, such as the operations system controlling US commercial air traffic. Others, like fire or police operations, must be prepared to deal with long periods of quiet interspersed with intense activity. In the face of all of these, the systems must be designed and executed as a routine, production oriented set of tasks.

Obviously, operations systems vary widely in complexity, required reliability and flexibility, and in many other ways. Different spacecraft and space payload operations will have varying levels of each. The Landsat Mission and typical weather satellites are designed to repeat the same or similar events daily for long periods; encounter missions, such as Voyager, are highly active only for brief periods which are separated by long, sometimes uneventful transit times known as 'cruise'. All operations systems, regardless of their intended mission, must also be prepared to deal with problems as they arise — anomalies can turn the simplest mission into the most complex, and its operations system must be prepared to act accordingly.

Most operations systems require high performance, and certainly many of them utilize new technology to help achieve it. Remote-sensing spaceborne missions not only use but require state-of-the-art performance in order to survive the naturally hostile environment that space offers. These requirements are imposed on many different fields: spacecraft, navigation, sensors, telemetry and other systems are all required to produce consistently, reliably and on schedule through the life of the operation. The technological problems are severe, as are the interpersonal communications and management issues that result when experts in these fields must work together and solve problems involving multiple disciplines.

Remote-sensing missions must be conceived, designed, built, tested and operated. Except for the last, each item on this list must be carried out in a reasonably studied manner. Operations, however, have a fundamentally different flavour, since they are intrinsically time-driven. Conception, design and construction must all be done with cost-efficiency, and of course they must be completed in a reasonable time period. Only the operations phase is driven by time on scales of minutes or seconds. Operations must deal with fast responses to anomalous events aboard spacecraft, which include some of the most complex machines developed by mankind. The process of response to anomalies may also need to consider characteristics of the target under study, where the local environment is only poorly understood. In these situations science and scientists — accustomed to operating in studied and slow ways so as to avoid making errors — are called on to make difficult decisions quickly, with incomplete data, and at great distances from the Earth. Nowhere was this scenario more obvious than in the Apollo and Viking landings, where only at the final moments before setting down on a new planetary body was it possible to discover whether the surface was too rough for the spacecraft to survive. It is an arena that challenges science, engineering, and even logic; and it is the mission operations system which must make it all happen.

Mission operations systems have in the past largely been designed one at a time — often with little experience base to draw on for either the system or design process. In the early days of space exploration, there was no such base. Later, missions and instrumentation became so individualized that the existing experience base was of little use — although it may be true that there has been commonality which was not taken advantage of until the multimission concepts of the late 1970s evolved. Having passed these periods, we have lately come to a time when it is financially and scientifically wise to profit from our experience by developing not only spacecraft and missions but also operations systems that are more generic in form. We hope this book will demonstrate that the operation of a remote-sensing mission can be well-designed and well-executed without being unduly expensive. The Magellan spacecraft nominal mission costs approximately US$100 000 a day to operate; the first Shuttle Imaging Radar (SIR-A) payload operation cost US$1000. Yet both were designed according to the general principles given here.

The book begins with an overview of the basics of remote-sensing missions and the definition of a few terms and concepts that are necessary in order to understand the descriptions that follow. Chapter 2 describes how one goes about the process of designing a mission operations system. This chapter starts with a description of generally accepted design phases and the design review system used by NASA to provide both management oversight and inter-team communication. It then moves to design documentation, which is central to the design process and forces the design to proceed from concept to requirements, through function and interface definition, to implementation and into the operations plans and procedures that detail the way the system works.

The next four chapters describe elements that must be present in any spacecraft-related mission operations system: management, uplink, downlink, and contingency response. These elements exist in some form in any size project: in a small, short-lived mission such as a Shuttle payload each element may consist of one or two people with a personal computer, whereas in a larger mission each element could use small armies of both people and computers. Each chapter shows the makeup of the element for both scales of activity. As part of each chapter, we have taken examples from missions we are familiar with, and show how these organized and operated the element.

Chapter 7 shows in more detail how the elements previously described fit together in one example, as we describe the Magellan mission to Venus. Some elements are modified to fit Magellan's particular objectives, but they are all here. Finally, the last chapter contains a description of recent developments that may change the look of mission operations systems in the future. Advances in such technologies as artificial intelligence, information systems, and data networking are described. From these examples it should be clear that the rapid changes which are occurring in all areas of technology will be needed and utilized by MOSs as soon as they are proven. The challenges of exploration and science are unending, and it is a requirement on the MOSs of the future to meet and to solve them.

Readers of this book should recognize that the authors have an acknowledged bias caused by their experiences, which have been almost exclusively in planetary and Shuttle-borne terrestrial missions. Although we have attempted to soften this bias with tours, discussions and other secondhand information, and with review by those with other experience, the bias undoubtedly persists in our examples and generalizations. We believe, however, that the principles expressed in this book can be applied to all types of scientific remote-sensing missions to effect an efficient, complete and cost-effective design.

Finally, we would like to acknowledge the help of the many operations experts that have contributed to this book. Operations is a combination of different talents, and we have depended on several sources to help us to describe them here. We would particularly like to thank Karen Moe of NASA's Goddard Space Flight Center, Raymond Buza of Martin Marietta Astronautics Group, Joan Horvath and Mike Jones of the Jet Propulsion Laboratory, and James S. Martin for careful review of early versions of the manuscript; Deborah Kristoff of JPL for her assistance in the preparation of Chapter 7; and Mike Davis of Hughes Communications, Inc., for information on commercial communications satellite operations organization.

Stephen D. Wall
Kenneth W. Ledbetter
January 1991

1

Basics of Remote-Sensing Missions

Remote sensing[1] is the act of acquiring scientific information using sensors that are not in direct contact with the object under study. Remote-sensing *mission operations*, in the context of this book, specifically refers to the control of one or more information gathering devices on board a vehicle in space and the associated operation of the vehicle systems in order to support information gathering. The function of a remote-sensing mission is to acquire remotely-sensed data from a *payload*, which might be an instrument or a group of instruments, aimed at a *target*, which may be an area on the surface of the Earth, a moon, planet, asteroid or other body, or may be a star field or other field of view containing objects about which information is desired. The data must then be transmitted to receiving stations on Earth (often referred to simply as the *ground*) for collection, processing, and interpretation.

In order to accomplish this primary objective of data acquisition, there is a secondary but vital objective, which is to operate and maintain the health of both the payload and the *platform* for the payload. The platform might be a vehicle such as the Space Shuttle, an orbiting space station, a free-flying spacecraft, or in certain applications, a robotic lander or rover on a planetary surface. A *free-flyer* is the typical concept of a spacecraft containing several remote-sensing instruments and their support subsystems. Instruments and support subsystems may or may not be grouped together as separately identified payloads. An idealized unmanned platform for several payloads is illustrated in Figure 1.1. Each payload can have several experiments with their associated support hardware, and each can contain their own support resources or draw on those of the centralized spacecraft subsystems.

If the platform is the Space Transportation System's Space Shuttle orbiter, the payload is generally mounted on a pallet in the Shuttle cargo bay and operates from the bay only during the flight when the bay doors are open. These missions are very short (typically 2 to 10 days) when compared with the dedicated planetary or free-flying Earth orbital missions which generally last several years. A more complex platform such as a space station will actually contain many payloads operating more or less independently, each having its own set of instruments. In many cases, each payload will provide its own subsystem resources. The planned lifetime of the space station Freedom is more than 10 years.

[1] Throughout this book, terms in italics are terms which are defined in the Glossary.

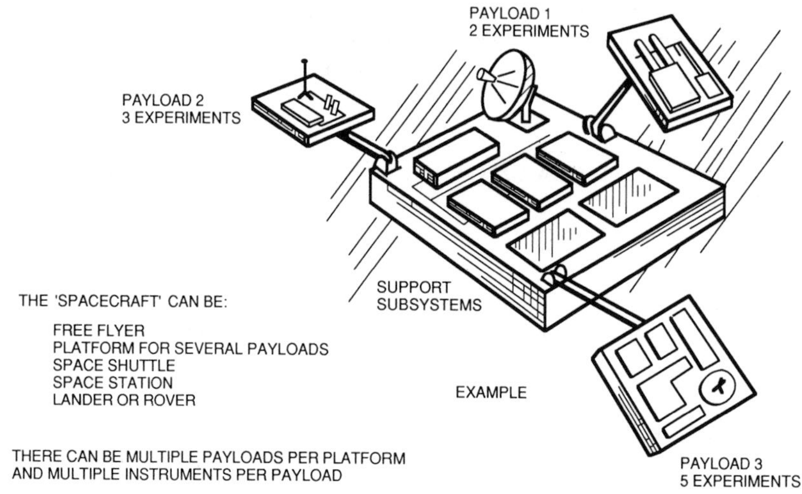

Figure 1.1 A multi-level platform definition.

For a mission operations system to maintain the health of the platform carrying the instruments, it must monitor and maintain each of several subsystems required for the spacecraft to correctly function. Maintaining proper operation of each platform subsystem, including keeping critical parameters within established bounds, will ensure that the subsystems can continue to provide required resources for the duration of the mission. However, resource limitations must be recognized so that the demand placed on the subsystems by the payloads will not exceed the platform's capacity to supply. Although it is not the purpose of this book to present spacecraft design, it is important in mission operations system design to have a basic understanding of the flight vehicle. Therefore, the following sections will discuss what constitutes the platform we desire to operate.

1.1 The platform

A platform is a flight vehicle system composed of subsystems. Like any system, the platform does not properly function unless all of its individual subsystems are properly functioning. On a free-flying spacecraft, all flight vehicle functions must be contained within the on-board subsystems. The platform must be autonomous. However, payloads hosted on the Shuttle or a space station may draw some of their resources directly from the host platform. Table 1.1 lists the required subsystems for a free-flyer in the first column (each of these will be addressed in detail in immediately following sections) and identifies the most likely location of these spacecraft functions for missions where the platform is not a free-flyer. All functions must be

Table 1.1 Location of subsystem functions for non-free-flyers

Subsystem	Space Shuttle	Space Station
attitude control	Shuttle	station
data handling	payload	payload
flight software	payload	payload
mechanisms	payload	payload
power	either	station
propulsion	Shuttle	station
telecommunications	either	either
thermal control	both	both

performed, but in those provided by the Shuttle or space station, the payload operations crew will be given the parameters and limitations of the providing subsystem and, in general, will not have to monitor its performance.

As an example of a free-flying platform, Figure 1.2 is a photograph of the Magellan spacecraft being prepared for pre-launch testing. This particular spacecraft will be described in more detail in Chapter 7, but just a quick glance will identify some of the principal subsystems that are the heart of its constitution: antennas for communication; solar panels for power (folded downward in this view); the thrusters and engines of the propulsion module for manoeuvres and attitude control; an optical telescope for finding reference stars (the small black circle in the very centre of the picture); blankets for thermal control; and structural bays to house flight computers, memories and other electronic components. Figure 1.3 is a photograph of a typical electronic flight component, this particular one a relay board *slice*, or removable unit consisting of printed circuit boards, electronic elements, wiring, and connector jacks.

The major function of the mission operations system engineers is to keep all these platform subsystems playing together in harmony, to allow the payload to perform its remote sensing function. Each of the subsystems in a typical platform is briefly described below (in sections 1.2 through 1.9) for future reference. Following this, in section 1.10, is a summary discussion of the instruments for remote sensing that can constitute the payload.

1.2 Data handling and storage

Flight computers are the brains of in-flight vehicle operation. They must decode and distribute commands received from ground stations directed to the on-board subsystems to instruct the spacecraft in its functions. These commands may be processed and issued to the subsystems in *real time* as soon as they are received or a time-tagged sequence of commands may be stored in memory for later execution as each command comes due. These com-

Figure 1.2 The Magellan spacecraft being readied for acoustics vibration testing in the laboratory. For the Magellan programme, there were no separate test versions of the spacecraft. The single flight article was tested and flown. Details of the spacecraft components will be described in Chapter 7. Photo courtesy of Martin Marietta Astronautics.

puters also prepare both the engineering health measurements from the subsystems as well as the collected science data from the instruments for insertion into the telemetry stream for transmission. Occasionally, the flight computer must take over operation of the vehicle from ground controllers, if an on-board anomaly causes a disruption in normal operations, performing a *safing* action to protect the platform or its payload. The components of a typical flight computer are shown in Figure 1.4. The slice on the left is the data processor assembly which handles computations, input and output. In the back are the banks of memory chips accessible to the processor. On the right is the power supply for the computer.

Since ground stations may not always be readily available to receive telemetry, most spacecraft contain a data storage device, such as a digital tape recorder, to hold data not immediately transmittable. Thus the data collec-

Figure 1.3 *A typical board component for a flight spacecraft. This example is a relay board consisting of a pair of circuit boards on a slice with several external connectors. Photo courtesy of Martin Marietta Astronautics.*

Figure 1.4 *An example of a typical flight computer. The memory chips are in the back, the power supply on the right, and the data processing assembly on the left. Photo courtesy of Martin Marietta Astronautics.*

tion path is divorced from the data formatting and transmission process, permitting data to be collected whenever the opportunity is there rather than when a ground receiving station is in sight. Onboard solid-state memory is sometimes used (and may replace tape recorders in future missions), but for any significant quantity of data, a tape recorder has been necessary. Figure

Figure 1.5 A typical flight tape recorder used by remote-sensing instruments for storing large quantities of data when that data cannot be returned to Earth in real time and must be buffered. Photo courtesy of Martin Marietta Astronautics.

1.5 is a photograph of a digital flight tape recorder. Note the two reels in the back. Two reels imply that the recorder must be able to reverse direction, or record and playback in both directions. Maintaining knowledge of tape position is a significant task for a mission operations flight team, so that such problems as unintentional overwriting of information or operating too close to an end-of-tape marker may be avoided.

1.3 Propulsion

The *propulsion subsystem* consists of the rocket engines (nozzles and thrusters) and fuel system (tanks, valves and plumbing) that allow a free-flying spacecraft to manoeuvre in space and modify its trajectory. A typical thruster module (sometimes known as a thruster cluster) is shown in Figure 1.6. Note that there are three distinctly different sizes of nozzles in this photograph. There are the two large engines that are obvious; one mid-size thruster (the small nozzle protruding from the cylinder); and three small tapered thrusters (two below the left engine). The different sizes allow different amounts of thrust to be applied over a specific time interval. In general, a propulsion

Figure 1.6 Propulsion thruster module containing six thrusters of three different nozzle sizes.

system is fuelled by a liquid propellant such as hydrazine, or, for less demanding applications, by a cold gas such as nitrogen. Occasionally a solid rocket may serve to perform a specific manoeuvre, such as for Magellan's Venus Orbit Insertion. This motor was shown at the bottom of Figure 1.2.

The in-flight operation of propulsion subsystem components is controlled by commands issued from the on-board computer initiated by a command sequence transmitted from the ground, or in some cases through signals received from on-board attitude sensors. Fuel lines can be opened or closed by pyrotechnic-initiated (explosive) valves and fuel line pressure can be maintained (to force fluid flow) by inert gas bladders inside fuel tanks that inflate as fuel is used. Activation and control of the various pyrotechnic devices, and the determination of their active state (disarmed/armed/fired) is a major task of the propulsion engineers on a mission operations team.

1.4 Power

Currently there are only three types of *electrical power subsystems* in common usage among remote-sensing spacecraft: solar, chemical and nuclear. Solar power is only practical for missions designed to operate close enough to the Sun to provide the needed power. In the past, planetary missions that have travelled outside of Mars' orbit have required more power than solar panels could deliver and therefore used nuclear sources instead. The Space Shuttle uses chemical power derived from combining hydrogen and oxygen, which

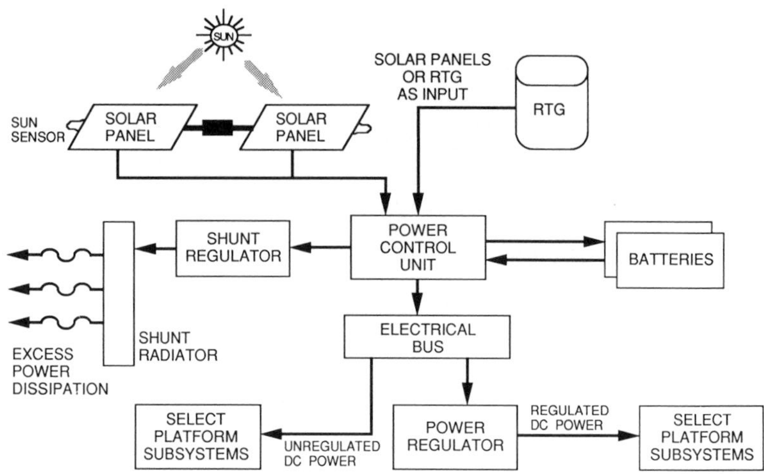

Figure 1.7 Major components of a spacecraft power subsystem.

produces water and energy. Chemical power alone, such as that provided by storage batteries, is only useful for short-duration missions, or in conjunction with other power sources.

The most common power source is a combination of solar panels and chemical batteries. In these subsystems, large panels containing photovoltaic cells convert incident sunlight into electrical energy used to keep batteries charged. Spacecraft functions, including instrument operations, draw their power from the batteries to ensure a constant voltage supply, especially important when the vehicle is on the shadowed side of the planet. A few short-duration Shuttle payloads may operate only on batteries for the length of their mission, but for any significant period of performance, batteries must be recharged. The second type of on-board power subsystem is the Radioisotope Thermoelectric Generator (RTG), which is a miniature nuclear power plant exemplified by the systems on the Voyager and Galileo spacecraft. RTGs can provide power directly to user devices without buffering by batteries, or as with solar panels, can be used to keep batteries charged.

Figure 1.7 shows diagrammatically a typical power subsystem for the type of remote sensing missions under study. Both solar panels and RTGs are illustrated, although only one of the two would actually be used on a given platform. The output from either the panels or nuclear power plant is directed through a power control unit to the rest of the spacecraft. The power control unit always ensures that the spacecraft subsystems are the first receivers of requested power via the electrical bus, which can either be regulated or unregulated direct current. Power left over will go into charging the batteries until they reach peak capacity, then excess power is dumped overboard by the shunt regulator and shunt radiators. If the solar panels are

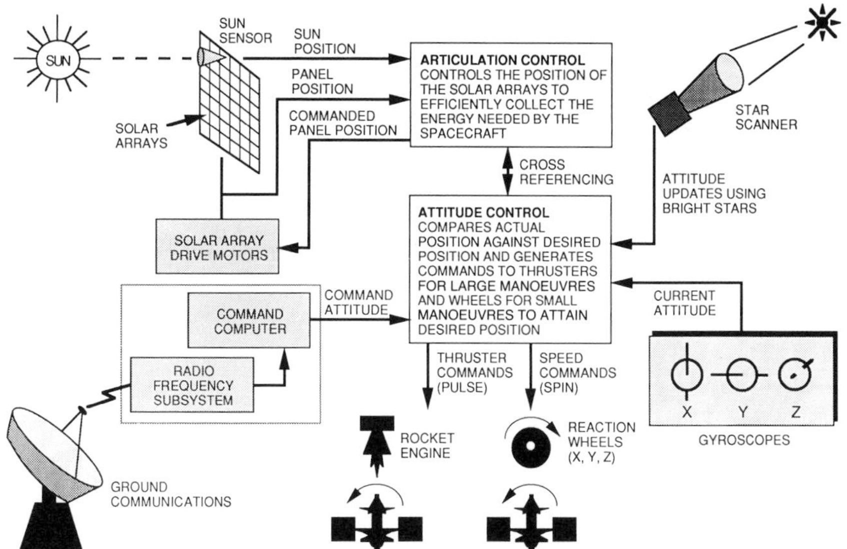

Figure 1.8 Attitude and articulation control subsystem for a 3-axis stabilized spacecraft with solar arrays.

not in sunlight, power will flow from the batteries back through the power control unit and out to the electrical bus and the subsystems.

1.5 Attitude control

The *attitude* (orientation in space) of the platform and the on-board knowledge of that orientation are crucial to nearly all remote-sensing activities. The two most common types of attitude control techniques are spin stabilization and three-axis stabilization. The former, such as for the Pioneer spacecraft, remains stabilized by spinning so that the entire spacecraft acts as a steady gyroscope. Three-axis spacecraft such as Voyager maintain a fixed orientation in space, except when manoeuvring, by viewing with optical sensors, the Sun and one or more bright reference stars. Deviations of measured positions from catalogued positions provide on-board corrections to attitude knowledge. Some Earth-orbiting spacecraft use Earth limb sensors instead of star trackers to determine their attitude.

Figure 1.8 presents a block diagram of an attitude and articulation control system for a 3-axis stabilized spacecraft with solar arrays. Articulation control is a positive feedback system that keeps the solar panels pointed toward the Sun by responding to sun sensor information and actual panel potentiometer positions with panel repositioning commands. Knowledge of platform

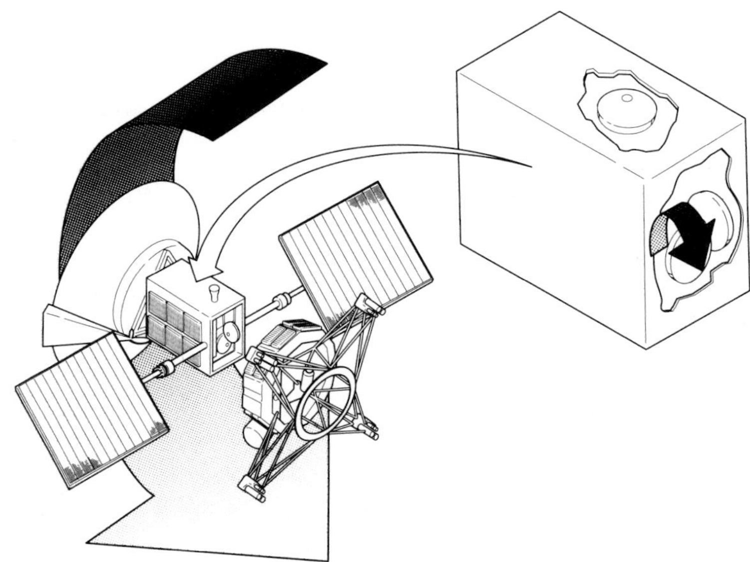

Figure 1.9 Reaction wheel control for spacecraft turning.

orientation is maintained by using gyroscopes and results of the star scans mentioned previously. Adjustments to attitude (small turns) can be accomplished under thruster control by the propulsion subsystem or, for selected spacecraft, through the use of gyroscope-monitored, spinning momentum wheels which can impart some of their rotational momentum to turn the spacecraft.

Momentum wheels have the advantage of being able to store rotational energy and release it back to the spacecraft as needed. For example, Magellan makes many turns toward and away from Venus as it collects radar imaging data and telemeters them back to Earth using the same antenna. A turn to Earth might be made by applying power to speed up a momentum wheel, as illustrated in Figure 1.9, and forcing the spacecraft to turn in the opposite direction by application of Newton's third law. The turn to Venus is then made by slowing the wheels and returning the spacecraft to its original attitude. At first glance, this system could work on its own forever, but because outside forces such as solar pressure and gravity gradient effects tend to apply rotational forces to a spacecraft, the momentum wheels must be used to absorb these disturbance forces as well. As a result of the accumulation of these disturbances, the wheels steadily build up speed, saturate and become useless in controlling the vehicle. Therefore, the hydrazine thrusters are used to slow them down when the wheels near their maximum safe speed, in a activity called *wheel desaturation*.

1.6 *Thermal control*

On the exposed surfaces of a spacecraft in deep space, the temperature differential between the side in direct sunlight and the side opposite the Sun may reach several hundred degrees centigrade. In general, protection of sensitive spacecraft components from these extreme temperatures is the function of multilayer thermal blankets, radiators, special reflective paint, louvers which can open and release internal heat, and thermostatically controlled electric heaters to warm up components that are becoming too cold. The external surfaces of some spacecraft are covered with *optical solar reflectors*, which are flat mirrors that reflect visible and ultraviolet light and emit infrared heat. The attitude of the spacecraft with respect to the impinging sunlight is usually the largest contributor to a spacecraft's thermal condition (with the possible exception of aerodynamic heating effects for those vehicles that must enter an atmosphere). Component temperatures, telemetered from strategically placed on-board thermistors, are carefully monitored by spacecraft ground controllers. Thermal considerations can significantly affect planning for upcoming mission activities.

1.7 *Telecommunications*

The radio (or *telecommunications*) subsystem provides the means of communicating with the spacecraft and its instruments both for command receipt and for telemetry transmission. It consists of one or more antennas, a receiver, transmitter, amplifier, and various signal generators to allow the spacecraft to communicate with the ground antenna at the proper frequency and data rate. Another use of the radio subsystem is for radio science, which will be discussed later in this chapter.

The principal function of the radio subsystem is to support the establishment and maintenance of the telecommunications link between the flight vehicle and the ground. While important for all spacecraft, this function is especially critical for planetary spacecraft where the link spans large distances and the signal strength is sometimes measured in billionths of a watt. Large antenna systems are required on the ground to 'listen' to the faint signals from space (Figure 1.10). The flight system must work in concert with the ground system. The first step is to synchronize the frequency of transmitted signals to the frequency at which the receiver is set. Note that there is a receiver and transmitter both on the spacecraft and on the ground, so this synchronization can be either one-way (establishing a telemetry link only) or two-way (establishing a coordinated command and telemetry link). Since transmitter and receiver are usually moving with respect to each other, the signal is frequency-shifted by the *Doppler effect*. Transmitter frequencies also vary with temperature and thus are not precisely predictable. The process of discovering a spacecraft's natural signal frequency and then matching it with

Figure 1.10 The Deep Space Network's 70m stations. In the upper left is the station near Canberra, Australia; upper right is near Madrid, Spain; and the lower image is near Barstow in California's Mojave Desert. Such large antennas are essential to maintaining contact with planetary spacecraft. Photo courtesy of Jet Propulsion Laboratory.

receiver tuning is called 'locking up' and can be quite difficult. Once locked, commands are transmitted to a receiver on-board and after the remotely sensed data are collected, they are sent from the on-board transmitter as a telemetry stream to the receiver on the ground.

1.8 Mechanisms

Mechanisms are electromechanical devices containing gears, motors, arms and actuators, that permit portions of a spacecraft to move independently from the rest. Solar panels, for example, must be free to articulate so that the panels remain as close to normal to the radiation from the Sun as possible for maximum solar power input. Optical instruments are sometimes placed on *scan platforms*, articulatable supports which can be rotated in one or more axes independent of the spacecraft's motion, to permit pointing the instruments at a target. Parts that move are always vulnerable to problems, therefore a flight team must monitor their status carefully and have plans ready in case they malfunction.

1.9 Flight software

As computer memories have grown and in-flight capabilities increased, more and more control of the spacecraft's functions is being moved from the ground controllers to flight software. On the newer generation missions, software handles not only the operation of the spacecraft subsystems and the collection, formatting and transmission of data, but maintains attitude control through on-board attitude knowledge determination. For planetary missions, where delays in communication due to light travel time may reach hours, flight software generally incorporates an on-board emergency life-saving process called *fault protection.* Software algorithms, formed into routines which run in the computer background, monitor spacecraft configuration states and critical parameters such as temperature, altitude or time since last communication with the ground. If these parameters exceed programmed limits, an appropriate fault protection algorithm is activated to take a predefined action. These algorithms range from basic functions such as turning off non-critical equipment to complex manoeuvres which autonomously re-point the spacecraft antennas to Earth and put out a call for help.

1.10 Remote-sensing instruments

There is a large variety in the types of remote-sensing instruments that have been flown on past missions. The sensor employed depends on the kind of data desired. The choice of sensor also depends on the dimensions, distance, spectral characteristics and other parameters of the target under study, as well as on more practical considerations such as cost, size and weight.

The three basic classes of information about targets are spatial, spectral, and intensity (Elachi, 1987). *Spatial information* tells us how different parts of a target relate to each other physically, which may lead to age relationships, methods of emplacement, or definition of physical mechanisms at work. Meteorological and geophysical information can also be derived from spatial data, for example from imaging of solar system bodies. *Spectral information* reveals the fine structure of solid targets at the scale of the radiation wavelength, or compositional information from measurements of

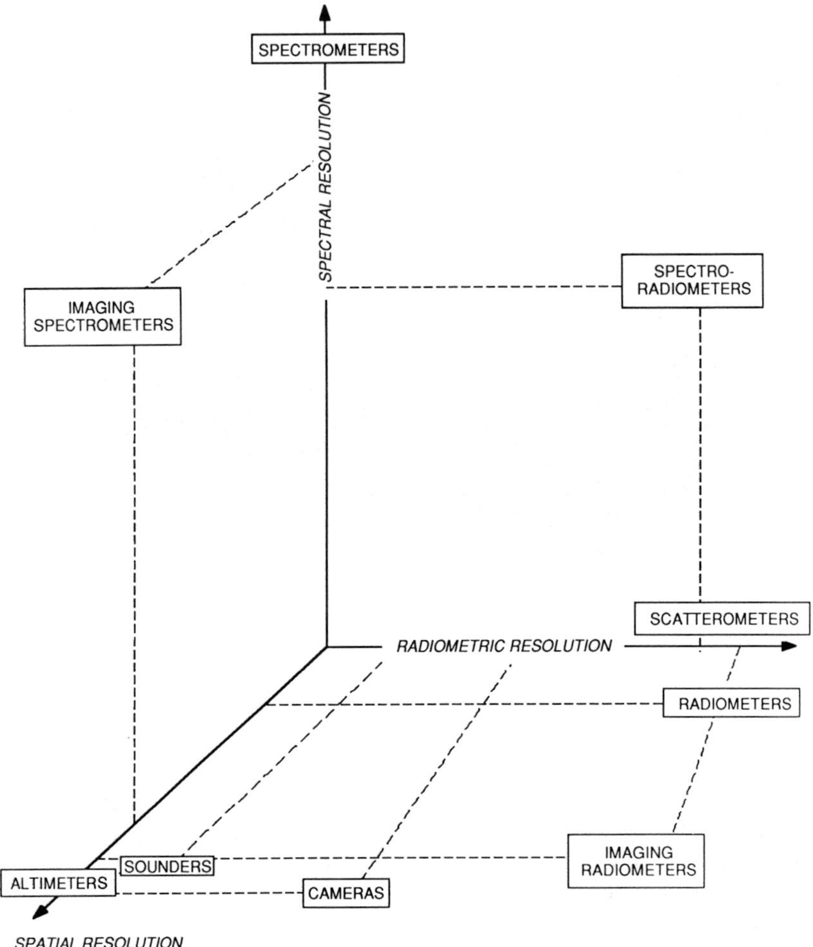

Figure 1.11 Relationship of sensor data type to data category. The three axes represent the three basic types of remote-sensing data, and various examples of instrumentation are shown.

radiation passed through gaseous material, as in stellar atmospheres. *Intensity information* measures the amount of radiation received. Figure 1.11 relates the different types of sensors to be discussed below with the basic classes of data they are designed to acquire.

For spatial information gathering purposes, two-dimensional imagers, loosely called 'cameras', are used. Cameras can provide both a large, synoptic view of a target or, with different settings or lenses, provide high-resolution data of smaller areas. They can sense radiation in the ultraviolet (wavelength

Figure 1.12 The Voyager narrow-angle camera. Photo courtesy of Jet Propulsion Laboratory (P-321686BC).

range 0.2 to 0.4 μm), visible (0.4 to 0.7 μm), infrared (0.7 to 10^3 μm), or microwave (10^3 to 10^6 μm). Each wavelength region measures a different property of the target. Different choices of bandwidths are available as well, although camera systems generally prefer better spatial discrimination at the expense of narrow spectral bands.

For both scientific and public relations purposes, imaging systems have been the most common spatial instruments on remote-sensing platforms. A photograph of one of the two cameras flown on the Voyager spacecraft is shown in Figure 1.12. The first imaging systems were outgrowths of television cameras. Even today optical imagers do not differ greatly from commercial TV systems, although resolution demands are much greater and digital systems have largely replaced analog. The original vidicon tubes have been replaced with two-dimensional arrays of solid-state sensors such as charge-coupled devices. For lander and rover missions, optical-mechanical scanners which employ a set of point detectors combined with two-dimensional mechanical scanning systems have been widely used because of their lower power and thermal resource requirements. These devices can also be adapted to many different output data rates without memory buffering.

Another form of spatial sensor is the altimeter, which emits a pulse of radiation and measures the time taken for the pulse to return to the sensor from the target. Such instruments are useful for mapping the topography of planets in conjunction with gravitational studies. Similar devices called sounders are used to measure subsurface structures or layering, as for example in polar ice caps.

Spectrometers take the opposite preference, spectral information at the expense of spatial. Targets such as atmospheres (planetary or stellar) have lower inherent spatial detail and require more spectral information to deduce chemical composition and chemical changes. Terrestrial pollution studies employ limb-scanning spectrometers, for example, where natural sunlight is observed after the atmosphere has absorbed some wavelengths. Because of the geometry, spatial resolution is poor, but the long absorption path lengths allow more quantitative information about minor atmospheric constituents. As a compromise between spatial and spectral requirements, imaging spectrometers do well at producing large amounts of both types of information with accompanying large data rates.

Radiometric information relates to the amount of energy returned from a surface. Active radiometric sensors radiate their own energy and measure the percentage of that energy that is returned. Most radiometers are passive sensors which measure either the natural radiation of a target or the amount of natural energy (e.g., sunlight) reflected. Radiometric data are useful for measuring quantitative amounts of whatever parameter is of interest, and at infrared or microwave wavelengths can be used to infer the temperature of a surface. Special forms of radiometers include scatterometers, which measure in a narrow spectral bandwidth the amount of radiation scattered back to the sensor as a function of angle, and polarimeters, which divide and measure radiation in its basic polarization states.

The measurement of electromagnetic radiation is the most common form of remote sensing, but it is not the only one. Magnetometers on both Earth-orbiting and planetary spacecraft are used to detect and measure magnetic fields. Charged particle detectors have been widely used on a variety of spacecraft ever since the early Explorers detected the van Allen radiation belts around Earth. Sensors that record the presence of basic particles in interplanetary space are used in planetary missions. Detectors of high energy protons (and other particles) emitted by the Sun can provide much information about our star.

For a lander, rover, or penetrator platform, there is a wide variety of instruments available. At least in theory, any instrument that is employed terrestrially in either a contact or non-contact mode can be put on to such a platform. In practice, there are severe limitations on power, weight, space and thermal environments that are allowable on any platform that must survive entry through an atmosphere. The constraints on instruments placed aboard hard landers and penetrators are even more stringent. Still, a great variety of instrumentation has been successfully used on all of these platforms. Soft landers have employed mass spectrometers, a variety of cameras, seismometers, and meteorological instruments to measure air temperature, wind speed and direction. Those that have soil sampling arms have used simple devices such as magnets, sieves, and even painted grids to analyse collected samples. X-ray diffraction experiments, and even specialized robotic biology experiments, have been developed based on their terrestrial laboratory analogues.

Some types of remote sensing do not require their own instruments but use engineering subsystems of the spacecraft. The field of radio science has evolved with planetary missions, where the spacecraft telecommunications subsystem is used to direct radio frequency radiation through a planet's atmosphere as the planet occults the telemetry path to Earth. As the radio signal fades, information about the constitution of the atmosphere is obtained. Finally, gravity investigations use the normal spacecraft positional tracking data to discover how the actual path taken deviates from that expected. From these differences it is possible to derive unexpected deviations in the gravity fields of planets.

Not only have remote-sensing instruments themselves changed over the last three decades, but the analysis products resulting from their use have also changed. If we consider cameras that return images, discussed above as the most common spatial instruments on remote-sensing spacecraft, the range has been wide. The data streams from cameras on early Mariner Mars spacecraft in the 1960s were dumped as raw numbers on computer paper. To obtain the image product, different ranges of numbers were hand-coloured with different coloured crayons, as seen in Plate I. Later, processing became more sophisticated and the digital image data were converted into film from which normal photographic prints could be made. Then it was discovered that useful analysis could be accomplished from photographic products made from non-visual data. Plate II presents three images of the same area of Death Valley, California none of which are normal optical images. The image on the left, Plate II(a), is a visual and near-infrared image taken by the Landsat Thematic Mapper. Plate II(b) in the centre is a thermal infrared image taken by the Thermal Infrared Mapping Spectrometer flown in a high altitude aircraft, and Plate II(c) is an image reconstructed from *synthetic aperture radar* (SAR) data.

The advent of SAR imagery has provided the planetary science world with a new tool. While the vidicon cameras flown on other missions, such as Voyager, could return detailed visual images of most of the solar system, as in the colourful picture in Plate III of the volcanic structure Pele on the Jovian moon Io, they were not useful in penetrating the thick clouds of Venus. It was left for the Magellan SAR to obtain the images of the surface of Venus, and it did so in spectacular fashion as evidenced by images such as that of another volcanic structure, Sacajawea, in Plate IV. The application of SAR imagery in the future will be wide, ranging from imaging the subsurface under the Sahara desert sand to unveiling the surface of Saturn's moon Titan.

Figure 1.13 The Magellan Spacecraft Team's remote mission support area in Denver, Colorado, 1500 km from the main control centre in California. The team processes live spacecraft telemetry and builds command files. Photo courtesy of Martin Marietta Astronautics.

1.11 The mission operations system

Now that we have a rudimentary understanding of what is in flight that must be controlled from the mission operations system, we can turn our attention to the ground. The heart of the mission operations system is the platform and payload operations control centres, those locations where the control of in-flight activity is maintained. This centre may be a highly visible, formal operations centre, as in Plate V, or a more subdued working area consisting of desks and workstation networks, as in Figure 1.13. In either case, this is where the major real-time decisions concerning the mission are made.

There is more to an MOS than the control centres, however. Figure 1.14 is a simplistic illustration of the principal components of the full system. The spacecraft transmits telemetry to a downlink station antenna on the Earth which reformats and relays the data to either the platform or payload control centre, which may or may not be the same. There the telemetry stream is separated into engineering telemetry, used for assessments of the health and

Plate I *Image construction on an early Mariner flight. This image was created by dumping raw telemetry data numbers on to computer printouts and colouring different ranges of numbers with different coloured crayons. Photo courtesy of Jet Propulsion Laboratory (P-37462).*

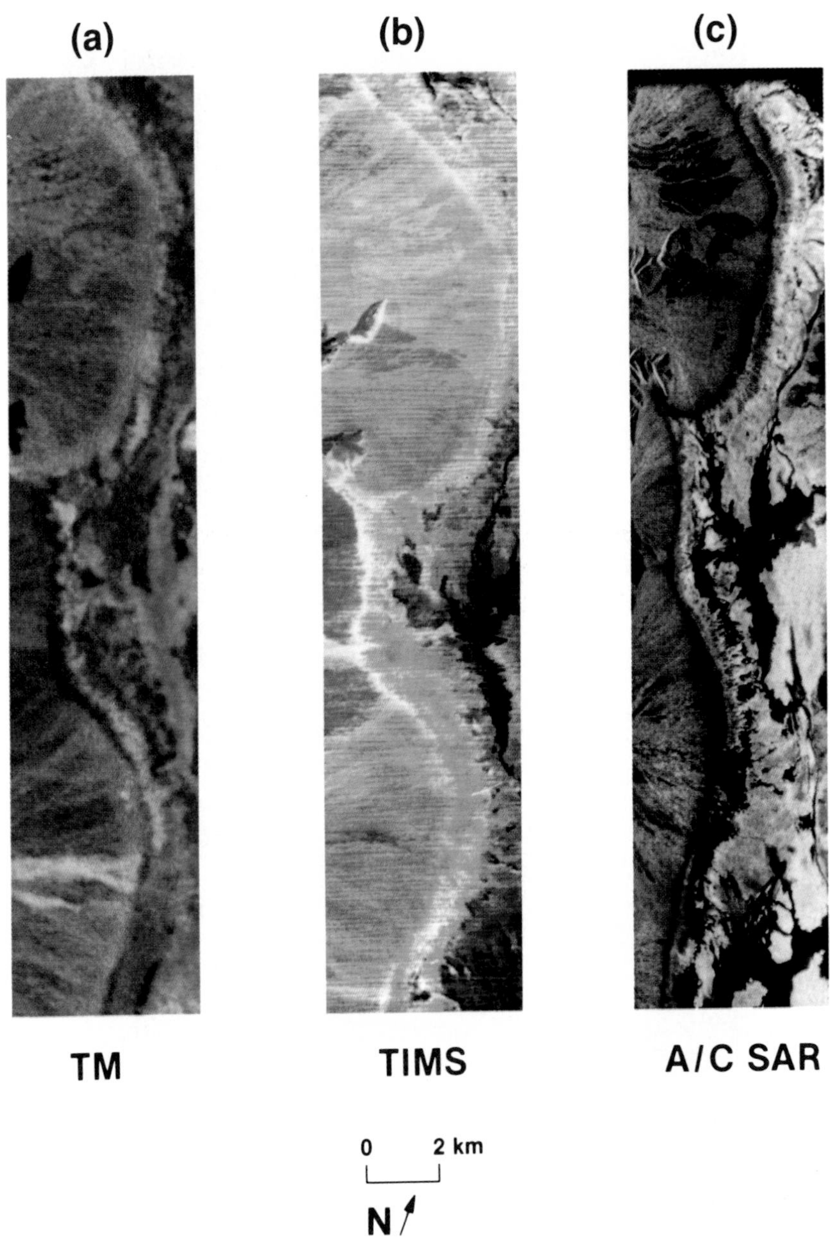

Plate II Multi-sensor images of Death Valley, California, an example of the sophistication that imagery has attained. The image on the left (a) is visual and near infra-red data from Landsat, while the image in the middle (b) was taken in thermal infra-red and the one on the right (c) was reconstructed from SAR data. Photo courtesy of Jet Propulsion Laboratory (P-30289BC).

Plate III Volcanic structure Pele on the Jovian moon Io, taken by Voyager's vidicon camera. The colour reconstruction is based on the use of multiple filters on board the spacecraft. Photo courtesy of Jet Propulsion Laboratory, California Institute of Technology (P-21226).

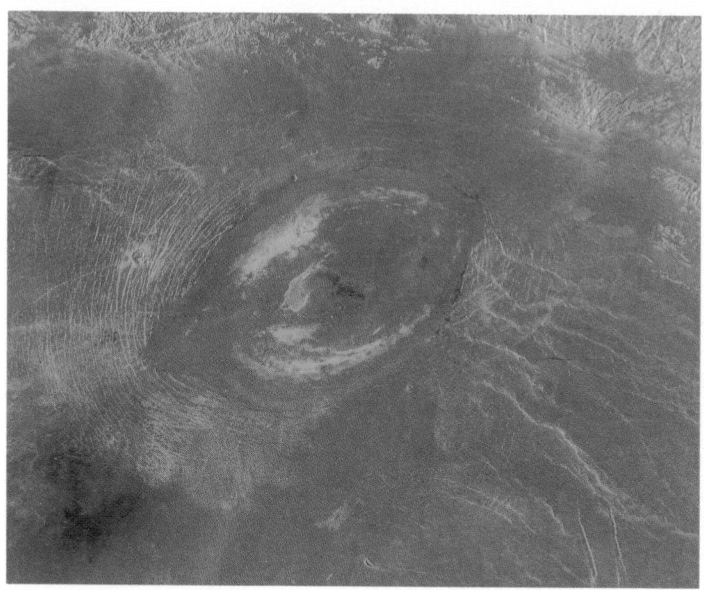

Plate IV Volcanic structure Sacajawea on the western plains of Ishtar Terra on Venus, reconstructed from SAR data taken by the Magellan spacecraft. Since Magellan's SAR only supplies one band of intensity information, the colour version required the addition of Venera lander colour data. Photo courtesy of Jet Propulsion Laboratory, California Institute of Technology (P-37137).

Plate V The Mission Operations Control Center in the Space Flight Operations Facility at the Jet Propulsion Laboratory — the control centre for many early lunar and planetary spacecraft. The centre is used now primarily for multi-mission operations. Photo courtesy of D. Connor.

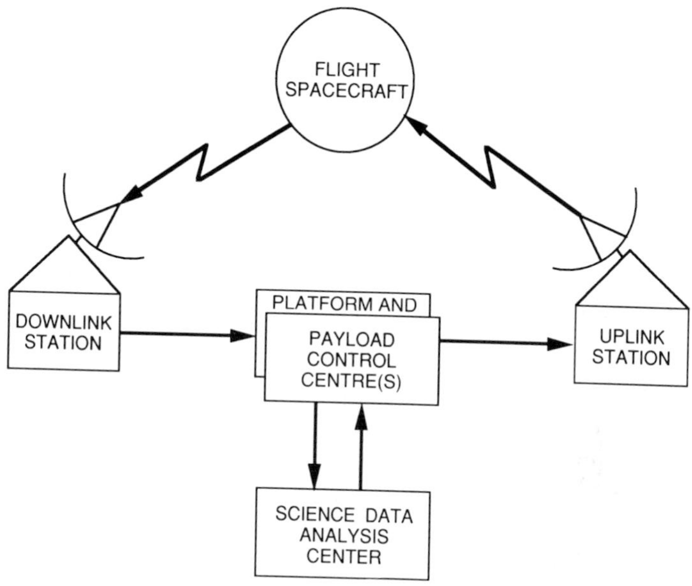

Figure 1.14 Simplified overview diagram of a mission operations system.

status of vehicle subsystems, and the collected remote-sensing data, which are sent on to the science data analysis centre for further processing and evaluation.

Command sequences are designed by payload controllers in response to the mission plan devised in advance by the science planners, then modified to the extent permissible by developments occurring after plan completion. These command sequences are sent from the payload or platform control centre to the *uplink station* (used to transmit information to the platform, which may be the same or a different antenna as the *downlink station*, the antenna used to receive information from the platform), where the commands are transmitted to the in-flight spacecraft and subsequently acted upon by the data-taking instruments and subsystems.

Figure 1.15 expands the top level diagram into a functional flow for the components of the ground data system. The exact configuration of such a diagram is highly mission dependent, but the basic elements are present in this figure to illustrate the required functions. Telemetry from the spacecraft is received, processed and monitored for alarms by the initial function. Part of this is done at the site of the downlink receiving station and part at the platform control centre. The telemetry stream is passed to another function that performs data handling, distribution and archiving into a permanent database for access. This function separates the data into their various types and distributes platform subsystem engineering telemetry to a spacecraft engineering data analysis function, instrument health and status telemetry to

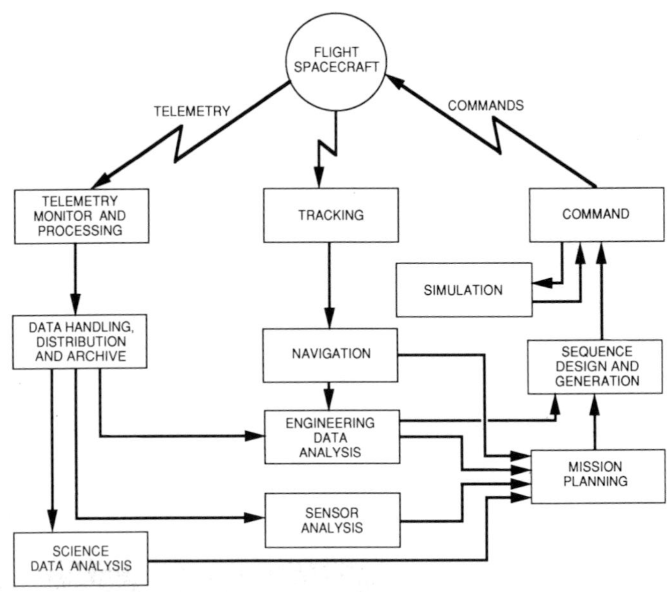

Figure 1.15 MOS ground data system flow diagram.

a sensor analysis function, and experiment remote-sensing data to a science data analysis function.

Each of the three analysis functions process and analyse their respective portions of the telemetry and make a determination of what to do next with the flight system. Commanding requirements are fed into the mission planning function which coordinates requests, resolves conflicts and builds the activity plan for the upcoming period. This plan is then input to the sequence design and generation functions to build, implement and verify the command sequences that are subsequently transmitted to the spacecraft by the commanding function. A spacecraft simulation function may be necessary to perform a final validation of the command load before transmission.

Somewhat standing alone, but nevertheless very important, are the tracking and navigation functions. Tracking is the function that collects positional and velocity information about the vehicle so that navigation can determine precisely where the spacecraft is and where it is going; in short, perform an accurate orbit determination. The collection of tracking data usually takes place remotely from the control centre at an antenna site, perhaps the uplink or downlink station, or at a multimission satellite tracking site. Navigation requires large number-crunching computers and is performed at the platform control centre. Output is fed into the mission planning process.

The above discussion is but a very brief introduction to mission operations systems. Each of these steps will be expanded in subsequent chapters. Note

Table 1.2 Key terms from Chapter 1

altimeter	power subsystem
articulation	radio science
attitude control subsystem	radio subsystem
downlink station	radiometer
fault protection	remote sensing
free-flyer	scan platform
intensity information	scatterometer
magnetometer	spatial information
mechanisms	spectral information
momentum wheels	spectrometer
one-way lock	target
particle detector	telecommunications
payload	thermal subsystem
platform	two-way lock
propulsion subsystem	uplink station

that the flow arrows in Figure 1.15 do not necessarily represent all of the possible interfaces between the functions that can exist.

1.12 Summary

In this chapter we have briefly described the basics of remote-sensing missions from the standpoint of the configuration and content of the in-flight vehicle. The various types of remote-sensing instruments that form the payloads aboard a platform have been summarized along with brief descriptions of the required platform subsystems that provide support to the sensing activity. Following this, an overview of mission operations systems was presented so that the reader can gain a perspective of what must occur on the ground to support the in-flight vehicle and experiments. Table 1.2 is a list of the key terms and concepts presented in this chapter. Armed with a basic knowledge of the in-flight system, we can now turn our attention to the details of the development and configuration of the mission operations system on the ground that must make the remote sensing occur according to a mission plan.

1.13 Exercises

(1) For a two-week mission consisting of an infrared astronomical telescope flown as a Shuttle orbiter payload bay experiment, define the necessary support subsystems and discuss which of them should be self-contained within the payload and which should be supplied by the Shuttle. Give reasons for each answer.

(2) Explain how an optical star scanning device might be used in a spin-stabilized spacecraft to establish attitude control reference. (Hint: This was done in the Pioneer Venus spacecraft.)

(3) Explain why a spacecraft's momentum wheels speed up rather than slow down with time and thus require regular desaturation.

(4) You are a mission operations engineer in charge of a 3-axis stabilized spacecraft in fixed attitude flight, collecting magnetic field data in very high Earth orbit and playing them back to ground stations for 15 min every hour. The on-board computer, currently located on the sunlit side of the spacecraft, is gradually approaching its high temperature limit as the solar angle on that side gradually approaches perpendicular. What are some of the actions you might take to cool the computer and continue the data collection mission?

(5) For a mission to the outer planets of the solar system, where round-trip light time telecommunications may reach several hours, what types of fault protection actions would you want to install in on-board flight software?

(6) Given data from Voyager's flyby of Saturn's large moon Titan showing a thick, opaque atmosphere with a surface pressure over twice Earth's, what types of remote-sensing instruments would you recommend for a spacecraft in a circular orbit around this interesting moon? Justify your selections.

(7) Explain why imaging spectrometers generally require larger amounts of data than spatial imaging systems or spectral systems alone.

(8) Give at least three reasons why the uplink station and the downlink station for a given operational mission may not be the same physical facility.

Reference

Elachi, C., 1987, *Physics and Techniques of Remote Sensing*, New York: John Wiley and Sons.

2

Design Methodology for Operations Systems

This chapter will describe the methods and techniques for designing mission operations systems (MOS) for spacecraft remote sensing within a context of standard programmatic steps and documentation used by both the National Aeronautics and Space Administration (NASA) and industry. It will introduce the phases of a programme and define the corresponding reviews that are typically required during each phase. Further, it will discuss the design steps from operations concept development through requirements identification and analysis, logical functional analysis, interface definition, physical environment definition, to operations plan and procedure development. The use of tools such as *N*-squared charts, activity timelines, interface definitions and scenarios, will be discussed. Throughout the chapter, practices which have added to the efficiency of design of mission operations systems will be mentioned.

In the past, the beginning of the MOS design occurred only after the spacecraft and mission designs were well established, due largely to the fact that mission operations occurs at the end of the programme. It has been difficult for a programme to consider downstream operations impacts when management attention is focused on an immediate spacecraft design or fabrication problem. This has led to innumerable difficulties in MOS design and implementation because decisions that impacted the ground system were made early by spacecraft and mission designers without proper input from MOS designers. An example of this was the way the information from the Viking lander seismometer was designed to be packaged into the downlink telemetry (see Chapter 5 for a detailed discussion of downlink telemetry systems). Knowing that if a marsquake were to occur, its signature could be adequately captured only by a high sampling rate, the seismology data frame was designed with the unique feature of expanding to a larger number of bytes if a quake were to trigger the instrument to flood the telemetry system with data. Telemetry frames of other instruments were fixed in length. This innovative idea may have served the experiment well during operations, but the design of the telemetry frame synchronization process on the ground was made unnecessarily complicated because of the variable length of this single frame type.

Lately, however, there has been progress toward getting the MOS designers involved in the process of spacecraft and mission design from the beginning. This chapter will show how a key document, called the operations

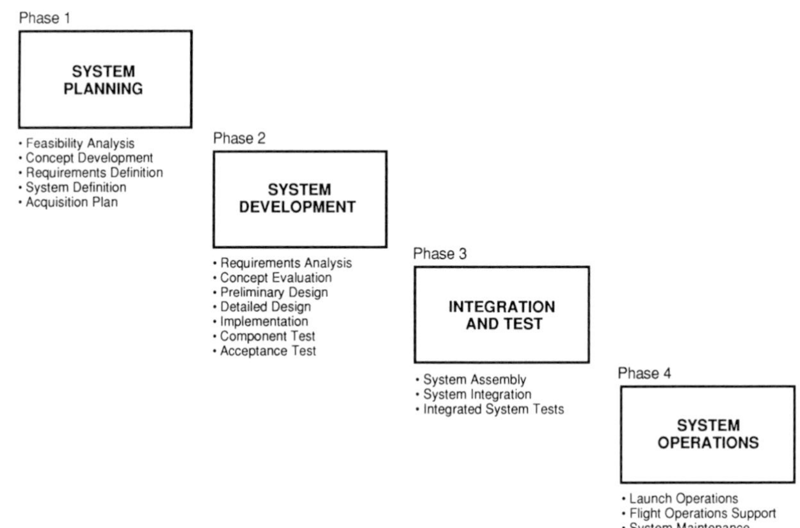

Figure 2.1 Phases of a system life cycle.

concept, can make this happen. Proper development and use of the operations concept document can result in the designs of the combined flight and ground system yielding enhanced operability, and thereby produce increased flexibility for less cost.

2.1 Programme phases and reviews

There are four principal phases of a programme's life cycle, as illustrated in Figure 2.1. They are: (1) *system planning*, which comprises feasibility analysis, system concept development, requirements definition, and the development of a system acquisition plan; (2) *system development*, which includes requirements analysis, concept evaluation, preliminary and detailed design, implementation, and test activities through acceptance testing; (3) *system integration and system test*, which encompasses the integration of systems into the overall mission operations system and complete testing of the integrated systems against all the specified requirements; and (4) *system operations*, which consists of flight operations support, beginning with launch operations.

In spite of the agony of extra work they cause for the system designers, formal reviews of the progress of the project before a board of peers and experts are essential to achieve design excellence. These boards, consisting of non-project personnel, provide an unbiased analysis of design and implementation progress, and typically contain past MOS managers, science representatives, outside consultants, major contractor experts, and representatives

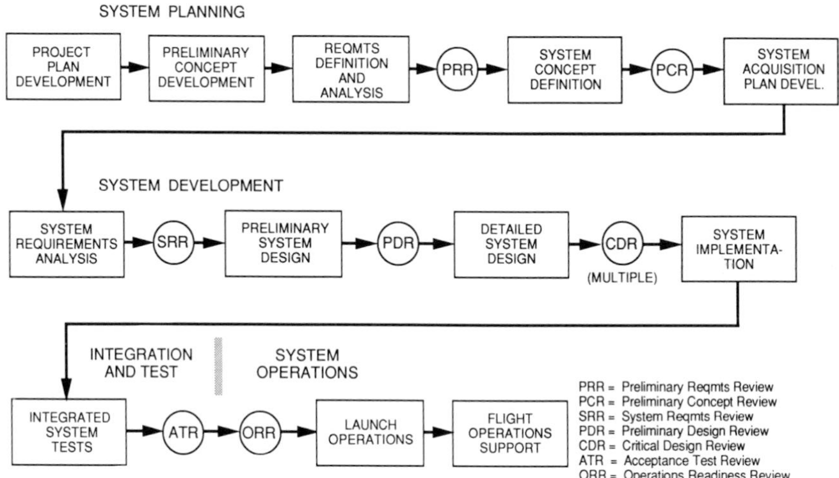

Figure 2.2 Programme phrases and reviews.

of the sponsoring agency. Figure 2.2 illustrates the timing of various reviews with respect to the programme phases. There is usually a Preliminary Requirements Review (PRR) after the initial requirements definition is complete, to assess if the requirements are well enough defined to allow effort to proceed. Historically, more projects have got into serious difficulties due to incomplete or ambiguously defined requirements than any other cause. There is sometimes a Preliminary Concept Review (PCR) after the initial definition of the system concept. These reviews occur during the system planning phase.

During the system development phase, a System Requirements Review (SRR) and the resolution of action items that follow will finalize the requirements and permit design to begin. After preliminary system design is complete, a Preliminary Design Review (PDR) is generally held. The preliminary design must be approved before detailed design can begin. When the design is complete, the Critical Design Review (CDR) is the milestone preceding procurement of hardware and start of actual implementation. In practice, multiple CDRs are usually held for various components of the system, because when delving into design details much more review time is required and different review board expertise is needed.

The above process is idealized, because much of the time the design has begun before the requirements are finalized. While this is not desirable, managers often must bow to the pressures of time and schedule. The readiness of different components for the start of design at varying times begs for beginning design before complete definition of system requirements is achieved. However, minimizing the amount of design accomplished before finalizing of requirements will lower the ultimate cost of the system through eliminating redesign. A prime function of the system requirements review is to evaluate readiness for the design to begin.

During the integration and test phase, there may be many reviews of specific activities to ensure the implementation and testing is proceeding according to plan. Components within the mission operations system, consisting of hardware, software or a combination of both, are incrementally delivered, after each has completed its own series of testing, including acceptance testing by the component end user. However, when the fully developed mission operations system is ready to be delivered to operations, an Acceptance Test Review (ATR) is accomplished. The ATR follows the acceptance testing of the system which has tested against the original requirements and against the operations concept. Once this gate is successfully passed, the Operations Readiness Review (ORR) is held to verify that the entire mission operations system, including the people who will operate it, is ready for operations to begin. This usually follows several training exercises, utilizing the system's hardware and software in simulated operational environments, to train the operations personnel.

For small projects where costs are limited and the complexity of flight components and ground systems are not as great, the reviews may also be less expansive in scope. The preliminary concept may be reviewed at the same time as the preliminary requirements, effectively combining the PRR and PCR. If the programme is small enough, the PRR, PCR and SRR may be combined into one single requirements review prior to design start. The PDR and CDR, however, should never be combined or eliminated no matter how small the programme. After the MOS is completed and tested, an Operations Readiness Review may contain the essence of an Acceptance Test Review. Although some reviews may be combined, it is important not to omit the actual step from the process, for in small projects as well as large, the efficiency of the resulting design depends on orderly execution of the design and implementation process.

2.2 Documentation

Documentation is a dirty word for some engineers and designers. It is surely much more fun to just jump in and start designing or building or coding, than to have to sit down and write down all you want to do before you do it. Nevertheless, complete and accurate documentation is essential to doing the job right the first time. There are several reasons for this. A person may find that the ideas and logic he had in his head are not quite so clear when he starts to write them down. The mechanics of the writing process itself helps to clarify the concepts and tie up loose ends. Documentation also permits review by peers before the commitment of time and financial resources to the next phase of the programme. Such a review will often uncover flaws not envisioned by the designer simply because he is too close to the design. This provides an opportunity to fix an incorrect or inefficient design before too much effort is wasted.

Getting the requirements (or the design specification, or the integration plan, or the test plan) for a specific system written down will also allow planners (or designers, or implementors, or testers) of related systems to see and understand how a given plan fits in with others prior to hardware or software development (or test) activity. This results in more accurate specification, better design and more complete testing of the interfaces between subsystems early enough to save unnecessary effort.

Requirements specifications are essential to establishing a common understanding between customer and contractor as to what the task must include and what the job must accomplish. Agreement to the requirements before proceeding to design is critical to efficient, cost effective, and on-time development of a system that will meet the customer's needs. In addition, requirements documents provide a source of test objectives for the programme test phase as well as providing a valuable archival reference during the remainder of the programme. Certain design documents are also useful for long-term archiving (see Chapter 5).

There are wide varieties of documents that can be written, depending upon the type of system to be developed and the particular desires of the contracting agency. When a hardware component is built, documents proceed from requirements specifications to design specifications to procurement description documents to as-built, end-item specification documents for the hardware configuration. For software, the flow is from requirements documents to design documents to release description documents to user's guides. Then, there are various kinds of plans and procedures and strategies. Since this book concerns the MOS design, we will address here only those that pertain to mission operations systems.

Figure 2.3 is a *document tree* for a typical spacecraft mission operations system. Not all of these documents are always required for every programme, nor are all the documents shown that a given programme may need to develop. Those documents on the far right are those generally developed during the design, build and test of the spacecraft but which are essential to retain and to maintain during operations. They do not fit specifically into the MOS document tree, but instead supplement it. To the left are those documents that are developed with the operations phase of the mission in mind, although, as we will see, they may be written very early in the life cycle of the program.

From the initial concept of operations both the mission requirements and the science requirements are developed. The MOS design requirements document will specify how the mission and science requirements will be converted to operations requirements to achieve implementation of the spacecraft operations control activity. Further down in the tree, requirements documents are followed, in general, by design documents, followed in turn by plans and procedures.

The principal operations document thread in the document tree is ex-

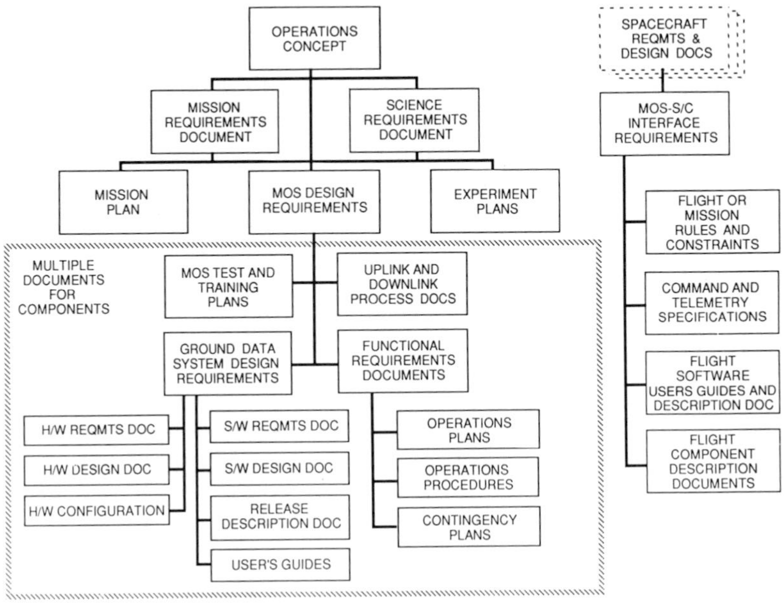

Figure 2.3 Typical document tree.

panded in Figure 2.4 to show the planning and development flow. An initial operations concept is first written, usually by the programme agency funding the mission. Its purpose is to provide information from which a contracting company or university can proceed to develop science requirements, mission requirements and the operations design requirements that support them. From operations design requirements that define what the system must do, functional requirements documents for various system components can be written as the operations concept matures. These are much more detailed, defining the required functions for specific components of the system, whether they are hardware, software or personnel. Reviews of this level of documentation often eliminate redundancy and can lead to decisions deleting components that are not really required.

In parallel with the system design effort, mission operations engineers begin to develop operations plans, using as an input the operations concept, the functional requirements documents and the continuing results of the system design effort. Much feedback and interface between the two parallel processes is necessary to optimize both. Finally, in order to implement the operating plans, operational procedures are devised which implement specific functions of the designed system step by step. These step-wise instructional procedures are useful for training personnel for the tasks they will later perform as well as recording actual performance during operations.

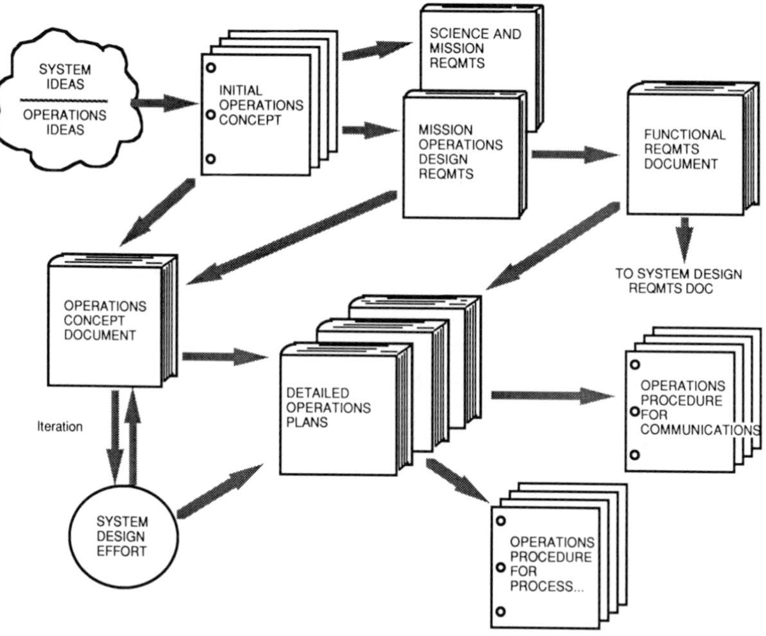

Figure 2.4 Operations planning and development flow.

2.3 Operations concept

An operations concept is an orderly collection of user-oriented ideas which show how the operations system should function to satisfy the mission and experiment objectives. Although it can exist without being documented, a written version ensures uniform dissemination of the concept. The assumption used here is that the operations concept is a summary document describing the collection of ideas that form the concept.

The operations concept is an evolving document, with a purpose and function that changes with the programme phases. Its initial and most important function is that it (1) drives the programme design toward one that will satisfy the mission and experiment objectives. Written early in the system planning phase, this initial version establishes and clarifies the intended operational approaches. As the design of the operations system is developed, the operations concept (2) guides the design engineers by shaping the definition of system requirements and keeping the focus on system operability. With additions and configuration-controlled revisions as necessary, the operations concept (3) becomes a summary description of the design at the end of the design phase, illustrating the way the operations system will be used to conduct mission operations.

The objectives of the operations concept and its uses are indicated by the following eight items, each allocated to one of the three document purposes introduced in the preceding paragraph (Ledbetter, 1984b).

As a design-driving document, the operations concept:

(1) Summarizes the initial objectives and constraints for both the mission and experiments. These basic goals are established by the project scientists and mission planners and become an input to the system planning phase.
(2) Documents early the intended operational approaches. Basic operational philosophy delineated early in the planning phase will simplify subsequent tasks and establish design requirements.
(3) Defines how users will operate and maintain the system. The domain of user activities is defined. System operations, system maintenance and required institutional support are all addressed early enough to influence the design requirements.

As a design-guiding document, the operations concept:

(4) Becomes the unifying document for the requirements analysis and design phases. By clearly defining the operational use of the system, it serves as a reference for designers, communicating the operations strategies to project personnel as system definition and design proceed. It also provides a test bed where design issues can be raised and resolved.
(5) Clarifies operational interfaces early. It identifies operational interfaces early enough in the programme to ensure a common understanding and sufficient definition, resulting in a more efficient implementation. Interface identification also defines the environment for the integration test programme by specifying which operational components must 'play together.'
(6) Provides a framework for trade and cost studies. By defining and prioritizing necessary system operational features, the operations concept will provide criteria for evaluating trade study and cost options.

As a design-description document, the operations concept:

(7) Provides input for generation of plans. It supplies information for various operations plans such as the Mission Data Plan (which describes the handling plan for down linked information) and the Experiment Operations Plan (which explains how the experimenters will operate their instruments). It also supplies test objectives for system integration test plans to ensure that the testing will prove the mission concept and that the design will meet established requirements.
(8) Summarizes intended mission operations. It always remains a concise, readable summary of the purpose and intended operation of the mission at any time during any phase, for either external or internal interests. It can be required reading for new personnel.

Although creation of the initial version of the operations concept involves experimenters and mission planners, the responsibility for generating and updating the full document falls to the Mission Operations Manager (MOM), a management position described in Chapter 3. Later, systems engineers will consult the operations concept for guidance to ensure that the system design will satisfy the operator's requirements.

The initial version of the operations concept is written after only the rudimentary mission objectives have been determined, and in fact must be formed in parallel with those early mission concepts. It is generated from conceptual ideas of how the system will be operationally used to satisfy mission objectives. At this stage, the contents are the objectives and constraints of both the mission and individual experiments, and the conceptual approaches to operations system activities. Once determined and agreed upon by the participants, the operational approaches that constitute this initial concept are placed under *configuration control*, meaning that they cannot be arbitrarily changed without consent and approval of all affected parties. This initial concept is the major source of operational requirements generated as an input to the systems design requirements document.

Once the initial set of mission operations design requirements are defined, the first version of a full operations concept document can be written. It is likely to have many incomplete sections for those areas where detail is dependent on a design selection. It will contain, however, the basic concept for operating the system as it is initially envisioned. It will define the intended uplink process (see Chapter 4) for planning, scheduling, generating, validating, and transmitting commands or sequences of commands, and the downlink process (see Chapter 5) for receiving, monitoring, separating, and processing the telemetry. These strategies are not final at this stage but will evolve with the design. Table 2.1 provides a sample outline for the content of a basic operations concept document.

To efficiently implement the operations concept suggested by the above methodology, several ground rules need to be imposed during development. Adherence to these rules, despite claims from project personnel that mission operations is too far in the future for immediate worry, may make the difference between success and failure of the operations concept as a useful tool (Ledbetter, 1984b).

(1) Obtain early agreement on basic operational approaches. It is probably more important that all the players agree on an operational approach than to have a perfect approach that lacks complete agreement.

(2) Keep the document concise yet comprehensive. It should be a summary of intended operations. It must cover all areas of operations necessary to accomplish the mission, but restricting details in each area to that essential to convey the message.

(3) Keep it updated. Revisions or additions, scheduled after major steps in

Table 2.1 Summary of operations concept document content

1. Introduction
   - scope and objectives of document
   - organizational roles and responsibilities
2. Summary
   - flight vehicle and ground system configuration
3. Objectives, constraints and approaches
   - mission objectives and constraints
   - experiment operations objectives
   - approaches to operations system activities
4. Operational requirements and functions
   - summary of operational requirements
   - functional definition and operations flows
   - definition of operational modes
5. Operational activities and scenarios
   - activity table and timelines
   - normal operational scenarios
6. Contingency operations
7. Operational interfaces
   - interfaces external to system
   - interfaces internal to system
   - human-computer interfaces
8. Operational hardware features
   - user equipment, operations console features
   - communications nets, data links
   - special facilities required
9. Personnel Component
   - operational positions and responsibilities
   - organization and support coverage
   - training concept
10. Reliability, maintainability and availability
    - system reliability and availability
    - maintenance and logistic concepts

the planning and design processes, are essential. To be useful, it must be current.

(4) Let it evolve with the design. Although the concept will levy operational requirements on the design, the concept should be allowed to change if the design effort indicates more efficient or less costly ways of implementing the requirements. The requirements themselves may also evolve with the programme.

(5) Keep its focus user-oriented. The focus must be on the eventual operation of the mission operations system to achieve desired mission objectives, and on the users who must operate it.

A well-written, complete, maintained operations concept document will contribute significantly to a well-designed, efficient and cost-effective mission operations system.

2.4 *Requirements definition*

In most programmes, the definition of requirements that will later govern the design usually begins to occur about the same time as the initial operations concept, with the formulation of top-level sets of both mission requirements and operational requirements. *Mission requirements* are high level statements of the goals and objectives of the mission itself — that is, what it is that the mission is required to achieve. In many cases, they are further delineated by science experiment requirements. Mission requirements may reflect such issues as the type of orbit necessary, the target characteristics to sense, the mission duration to achieve objectives, number of spacecraft contacts per day, spacecraft pointing accuracy, and the frequency range of an instrument.

Operational requirements are those requirements relating specifically to the methods of achieving the operational mission. In general, they define the scope of the ground activities within the mission operations system. They determine issues such as the number and type of ground antennas for spacecraft contact, the necessity for around-the-clock monitoring of vehicle activities, the types of computational activities that must occur, how experimenters will gain access to their data, and whether command sequences are validated with a simulator. At this level, the ground system can be treated as a series of black boxes (perhaps similar to Figure 1.15) where the operational requirements define the functionality of the box (what it needs to do) and its interfaces with other boxes, but do not delve into the details of how the correct product is achieved. However, to design the black boxes, each operational requirement must be decomposed into one or more functional requirements.

Functional requirements are those that govern the design of the system components to satisfy operational requirements by identifying the specifications for each function the component must accomplish. The primary emphasis of functional requirements is on how the system is implemented rather than on what the component does or the content of its operational product.

One additional type of requirement is necessary when defining the intended operation of the system being designed. *Performance requirements* are those which specify when functions or activities must be completed and their duration. They provide information such as the order in which tasks must be performed, the amount of time allowed for an activity to be accomplished, milestones that must precede or follow activities or otherwise be met, and resource constraints necessary, especially when computer operations are involved.

Requirements definition is one of the steps in the system planning phase, consisting of identifying and documenting the functional requirements for each component of the operational system. The product is usually one or more Functional Requirements Documents (FRD), which are written in directive requirements language such as: 'The telemetry processing module shall decommutate engineering telemetry into files containing . . .'. Sometimes there is more than one level of functional requirements document. The military versions of FRDs are divided into levels called the *A-, B- and C-specifications,* a series of requirements specifications with increasing amounts of detail. There is usually one A-spec for the entire operations system, several B-specs for the major first-level components (see Section 2.5 for a discussion of functional levels) and one C-spec for each subsystem or detailed component of the system.

At first glance, it may appear that the language used in requirements documents is overly stiff and formal, but it need not be. The language must, however, have two attributes to be effective. First, it must be clear and concise when it states a requirement. This is especially true when the requirement is likely to be expensive to implement; when resources become limited the wording of a requirement is often taken quite literally. For this reason, requirements document language sometimes approaches that used in legal contracts. Second, the language must clearly discriminate between descriptive and prescriptive statements, which explain the common use of the prescriptive word 'shall'. Note that to state 'the Sun will rise in the East' is descriptive and does not impose a requirement; to say, however, that an employee 'shall report to work when the Sun rises' is a clear imperative that the action occur.

Functional requirements for a mission operations system will usually have several different types of FRDs. In addition to hardware and software functional requirements for the components of the ground data system, there are FRDs for the teams of people that must support operations. This applies equally to the time-line intensive teams that control the spacecraft as well as those that are not so closely tied to a clock, for example, those that process image data products. Table 2.2 provides a sample outline for the content of a basic functional requirements document.

Designers use the functional requirements documents to develop their design, so it is critically important to come to a common and clear understanding of the functional requirements. Usually, programmes will require completed FRDs as a prerequisite to the SRR, and a go-ahead must be given at the SRR before design can begin. Once the requirements are identified and approved, preliminary system design begins and, in parallel, an analysis of the requirements is performed to verify whether or not they are complete and correct. One of the tools to accomplish this is functional analysis.

Table 2.2 Summary of functional requirements document content

1. Introduction
   - purpose and scope of document
2. Functional overview
   - summary of system objectives and intended operation
3. Operational environment
   - hardware and software physical restrictions
   - personnel or location restrictions
4. Functional requirements
   - detailed requirements specifications for all components of the system
   - hardware and software specifications
5. Interfaces
   - requirements for information transfer between this component and other components of the system
6. Traceability matrices
   - tables providing connections with higher level documents
7. Test requirements
   - requirements for testing the component functions
   - performance requirements

2.5 *Requirements evaluation through functional analysis*

Functional analysis is the study of a system through the examination of its elements as described by the system functions. These functions can be processes performed by hardware, software or by people. Functional analysis includes determination of the function's elements and the interfaces between them, including decomposition of each function into lower level functions. It also includes identification of the attributes of the system processes that occur within the function and the interfaces between decomposition components.

Functional analysis used in conjunction with an operations concept provides an organized representation of a system against which to specify and evaluate design requirements. The results form a structured framework (or model) upon which to evaluate both requirements and design. The model resulting from functional analysis can include hierarchy charts, data flow diagrams, *N*-squared charts, and interface attribute specifications, all of which will be discussed in subsequent sections.

A functional model can be either a physical or a logical model. Logical models are those that avoid physical orientations such as specific hardware or software, facilities or institutional elements. Physical models, obviously, are based around these existing physical elements. Logical functional models can be used more effectively than physical models when the task involves new design and implementation, because the logical model provides an unbiased framework for system specification. This ensures that as design specifications become detailed, physical biases do not cause requirements to be hidden. (Hyman and Ledbetter, 1984a)

Regardless of which type of model is used, model development requires a decomposition of functions into several levels. A single level function can almost always be broken down into multiple components. As an example, a simple function that states 'watch television' could be broken into:

1. Get out of the easy chair.
2. Retrieve the remote control from Junior.
3. Punch the ON button.
4. Select the channel to watch.
5. Return to the easy chair and sit down.

Each of these functions can be further decomposed. For example, step 4 can become:

4(a). Decide what channel to watch.
4(b). Inspect the remote control.
4(c). Push selected buttons with index finger.

Using a little imagination, the reader can probably select any of the second-level functions and perform a further decomposition. Notice that all functions at all levels contain a verb. A function is an activity that requires action to accomplish. While not absolutely required, using action verbs is a technique that helps develop complete and useful functional decompositions.

Once this decomposition process is complete, one method of presentation of the results is the *functional hierarchy chart*. An example functional hierarchy for a typical spacecraft mission operation is shown in Figure 2.5. This chart divides system operation into five generic first-level functions, specifically, to manage: (1) data capture, (2) data products, (3) experiment operations, (4) platform (spacecraft) operations, and (5) ground system operations. Notice in the figure that, except for the zeroth level, all functions also begin with verbs.

There are other ways to divide an operations system to produce a functional hierarchy. For example, one could divide the system into real-time operations, off-line operations, and support operations. Real-time operations would include such functions as telemetry capture and subsystems status monitoring. Off-line operations would consist of activities like experiment planning, schedule generation, or science data analysis, while support operations would encompass such activities as generation of reconstructed orbit profiles and providing communications services. Still another way to divide the system is chronologically, beginning with mission planning and continuing through spacecraft telecommunications contact operations to post contact data analysis.

No one method of division is inherently better than another. The goals of

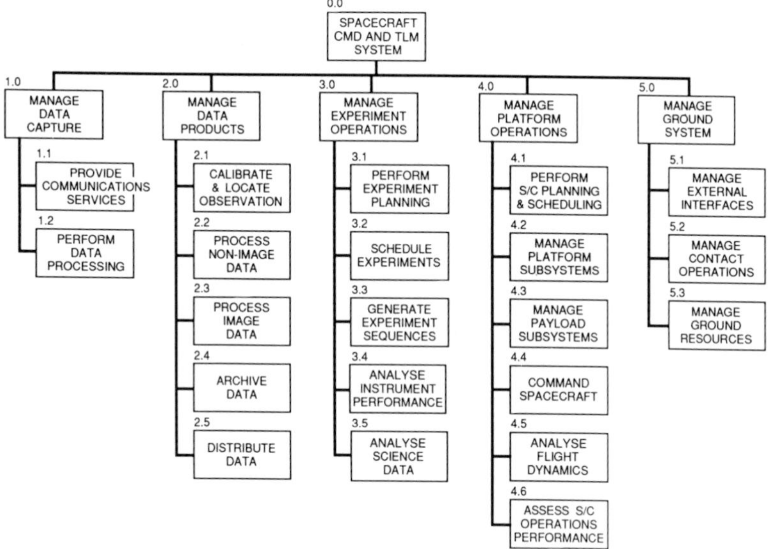

Figure 2.5 Logical model functional hierarchy.

system design need to be considered. For example, if a physically distributed operations system were an objective, it would be desirable to group functions expected to be performed by a designated set of people in one location, within the same functional grouping. Real-time instrument monitoring and long range experiment planning may both be performed by an experimenter at a university site located remotely from the platform operations control centre.

A top-level data flow diagram, for the functional hierarchy in Figure 2.5, is given in Figure 2.6. The primary function of a *data flow diagram* is to identify the data interfaces between the functions defined by the functional analysis. Like the functional hierarchy, the flow diagrams can continue to be expanded in greater and greater detail. Each ellipse in Figure 2.6 can be decomposed into its constituent functions and interfaces through several more levels. Ultimately, every required product interface can be identified through this method. Identifying and characterizing the interfaces between components of an operational system is a key to establishing an efficient, working design.

Figure 2.6 illustrates the standard conventions of a data flow diagram. The ellipses represent logical functions at a consistent level of functional decomposition. Lines between functions indicate information flow interfaces, single arrowhead lines represent one-way data flow and double-arrowed lines indicate data flowing in both directions. The parallel lines with words written in between (looking somewhat like the symbol for electrical capacitors) are

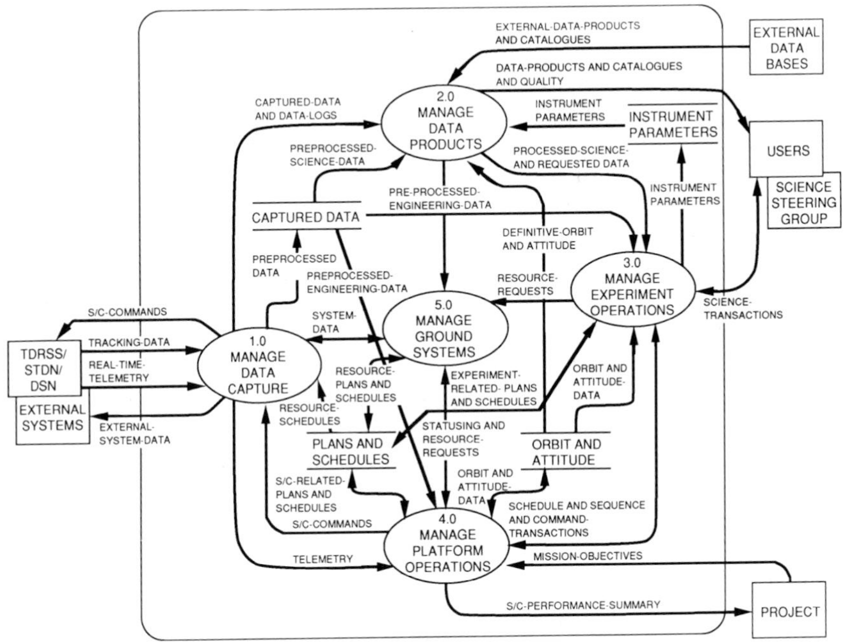

Figure 2.6 Command and telemetry data flow diagram.

read/write databases that store information for later retrieval by the same or another function. The large box around the figure describes the domain of the function we are decomposing (the 0.0 level in this case) and external interfaces are shown by lines entering and leaving this box. External interfacing functions are represented by rectangles outside of the domain under consideration, and they are not constrained to be indicated at the same functional level of the functions within the box.

One important consideration when designing data flow diagrams is to avoid drawing lines which represent organizational structure or communication lines that follow paths between organizational entities. While organizational structure is an important design aspect (and is treated separately in Chapter 3) lines of responsibility often run very differently from those of data flow. Communications lines are often numerous (in the extreme, going from every function to every other function) and can obscure the more significant interfaces. Another rule to follow is to consider omitting from data flow diagrams those functions which are themselves data paths. While they are necessary functions, computer networks and telecommunications equipment are examples of design implementation of the interface and should not be independently included, or the utility of the diagram is lost.

2.6 *Interface definition*

Once the system's functions have been clearly identified, and data flow diagrams have been generated to trace the flow of information through the system, interfaces between functions can be examined to investigate methods of designing and implementing the interface. The key to the ease with which a function can be implemented lies in the level of its interfaces with other functions. One that has a single data transmission interface is generally much easier to implement than one that has a great number of multilateral communications interfaces. However, a function with a single very complex interface may be more difficult to implement than one with ten simple interfaces. Therefore, both the number and the complexity of the interfaces between two functions must be considered (Hyman and Ledbetter, 1984a).

In general, the primary purpose of an interface is to transfer information from producer to recipient. A mission operations system can have three principal types of information transfer interfaces: electronic transfer; paper transfer; and voice transfer. Electronic data transfer can be real-time, which creates an interface involving the control of the environment where the automatic initiation, transmission and handling of the data is accomplished rapidly enough to affect the functioning of the environment at that time. Elapsed time and time of day are vital to the execution of a real-time interface because data flow must occur whenever events happen and must compete with all other simultaneous events. Electronic transfer can also be near real-time, where data, albeit in line with the operations flow, is delayed on the order of seconds to minutes, either by human intervention, software processing or hardware restrictions. Here, elapsed time and time of day are less important and there may be more flexibility to adjust transfer times to avoid conflicts with other data streams. Lastly, electronic transfer can be off-line to the operational control flow, where the data being transferred are not directly related to the current operations activity regardless of the link response time. This category includes interactive data retrieval or processing for purposes of planning or analysis related to a future operations flow activity.

Paper transfers also come in different categories of response time. These include request forms for various kinds of data acquisition and processing, printouts from data processing runs, and reports (and other products) from analysis, but in general they can be grouped into one category requiring human-to-human delivery. State-of-the-art operations systems strive to eliminate paper in human-to-human interfaces through the use of electronic mail. However, electronic mail systems do not yet provide all of the advantages of the form, in particular, the multi-part form. Approval signatures and supplemental information generated during task performance are difficult to append. A requirement (for example, a command transmission request) can be described on paper and a copy retained before sending it to the performing element. The sender's copy cannot be subsequently modified by the receiver. As the receiver accomplishes the requirement, he can write implementation details on the form, sign it and return a confirming copy to the person who requested it. The final form will contain requirement, implementation and signatures of both elements for a historical record of the action. There is no reason in principle that electronic means could not be devised to emulate such a system, where passwords are required to access selected parts of a file, but the authors are not aware of such a system in use.

Voice interfaces can be of two categories, normal telephone (dial-up) lines used for coordination of operational activities, and dedicated voice networks with multiple user access points. Dedicated voice networks are expensive, both to install and to maintain. There is no doubt that in large missions they are mandatory. They allow any individual within an MOS to instantly contact other positions without waiting for a connection to be made, which is particularly critical in anomalous situations requiring rapid response. They also permit a person to monitor many conversations for those relevant to his operational tasks. He can then be prepared to respond to occurrences before they filter down to his level of activity. Conversely, management can have visibility into operations far below their level without requiring subordinates to formally report to them. A common implementation of voice networks is to allow two types of access: listen only, and talk–listen. Typically, a given element might have listen-only access to networks several levels separated from his responsibility and talk-listen access to levels at and below his level.

One way to analyse interfaces between functions is to establish figures-of-merit based on interface characteristics. The number of interfaces is a sufficient figure-of-merit on its own, but complexity can derive from various characteristics. The following paragraphs define six interface characteristics that together define the complexity of an interface, that is, the 'strength' of the bond between the two functions that create the interface. The strength of the interface between two functions has serious implications if one of the functions is under consideration for remote operations in a distributed system.

1. Interface type. As described above, the types of operational interfaces are: (a) real-time electronic transfer; (b) near real-time electronic transfer; (c) off-line electronic transfer; (d) paper transfer; (e) telephone; and, (f) dedicated voice networks. Complexity generally increases as the information to be transferred comes closer to real-time.

2. Interactiveness (electronic interfaces only). Interfaces where the human interacts directly with the transfer process, by initiating the transmission and exercising options as to how the data are processed, stored or displayed, can be significantly different to implement than purely automatic processes.

3. Data rate (electronic interfaces only). In the design phase, the issue is not the assignment of a specific transfer data rate, but whether a high rate is known to be involved, as in telemetry rates from an Earth resources satellite in orbit, or so low that a commercial dial-up modem can be used.

4. Data volume. For electronic interfaces, data volume is considered to be defined by the size of the file transmitted in one communications link session. For non-electronic interfaces, it could be the thickness of the report required to convey the data. Like data rate, the assessment is qualitative during the design phase.

5. Frequency. In a spacecraft mission operations system many events occur on a regular schedule. There are three periods that seem to predominate: weekly activities, daily activities, and those that are associated with a spacecraft contact period. The contact period varies from once every 90 minutes for low Earth-orbiting vehicles to approximately eight hours for planetary spacecraft that contact each of the three deep space network stations once each day, to any period desired for geostationary satellites. The frequency is also higher if more than one user requires the same periodic data across the interface. For example, if a daily payload schedule is sent to each experimenter, the frequency of the interface between a central scheduling function and experiment operations is greater than daily.

6. Duration and link utilization (electronic interfaces only). Duration is the estimated length of time the communication link is maintained for one transmission. Link utilization is the percentage utilization of the link during the duration of the communication. Most off-line data transmissions occur very quickly with high link usage unless the volume of data is large. Most coordination communications are lengthy with low usage percentages. A real-time link is generally continuous but may not be fully utilized if it must be time shared with other activities. Interactive data links are often prone to low percentage utilizations due to time spent waiting for operator response.

These six characteristics, (and probably others) can be value weighted according to their significance to the specific operational system and summed into a figure of merit. This, together with the number of interfaces between the functions, can be evaluated to decide the merits of the implementation design.

One useful tool to display results of *interface analysis* is the N-squared chart, an example of which is shown in Figure 2.7. The various functions are shown along the diagonal of a rectangular grid consisting of N^2 intersections, where N is the number of functions under analysis. Each intersection represents the collective interfaces from one of the functions to another. (The reverse flow is represented by a different intersection.) Common convention dictates that the intersections to the right of the diagonal represent the interfaces from the upper function to the lower one, while those intersections to the left of the diagonal represents the return interfaces from the lower functions to the upper. That is, all interfaces follow a clockwise flow rule. The intersections themselves can contain different data, such as the sum of the number of interfaces between the two functions, a value representing the strength of the interface, words representing the nature of the interface, or a reference designator which points to a list of all the specific interfaces.

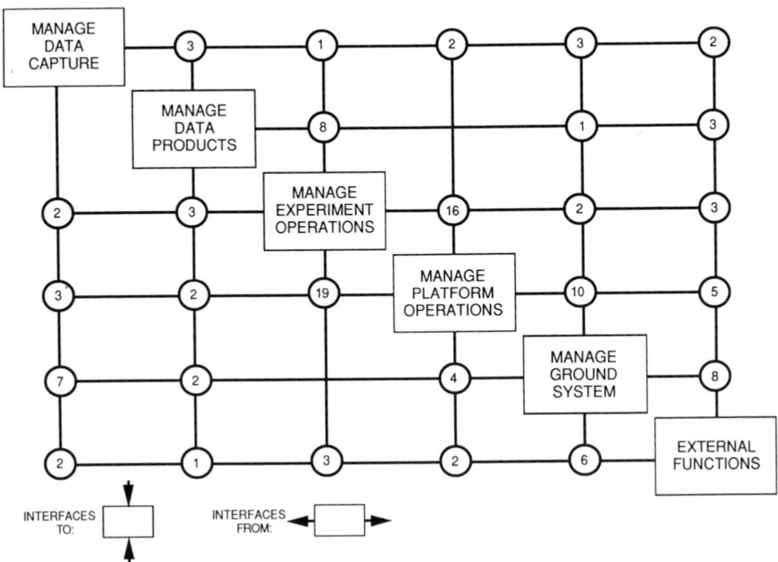

Figure 2.7 Example N-squared chart: number of interfaces.

2.7 Physical environment definition

The above discussion of generic mission operations systems is perfectly applicable for theoretical studies, but eventually the designer must apply the restrictions of the physical environment to the evolving design. The vast majority of spacecraft launched today will use many existing facilities, equipment and software which are likely to also be supporting other missions. This severely restricts the design in areas of specific interfaces with the external world. Only those programmes that build new ground control centres and new telecommunications systems along with new spacecraft, such as the Navstar global positioning system (GPS), can be free from such restrictions.

There are several existing telemetry, tracking and command (TT&C) stations of which the designer needs to be aware. For NASA's Earth-orbiting science satellites controlled from Goddard Space Flight Center (GSFC), the operations system must conform to either the Satellite Tracking and Data Network (STDN) stations or the Tracking and Data Relay Satellite System (TDRSS). Most current terrestrial missions use TDRSS as the prime system with STDN as a back-up. Space Shuttle flights also use TDRSS. For planetary probes, there is the Deep Space Network (DSN) at the Jet Propulsion Laboratory (JPL) consisting of three stations spaced approximately 120° apart around the globe. Air Force Space Command satellites use the Remote Tracking Station (AFRTS) network of eight stations worldwide. Commercial communications satellites have their own antennas and control centres, generally also used for multiple satellites.

The significance of these TT&C facilities for the mission operations system designer is that each one has a defined set of interfaces to which any new design must adhere. Command files must be formatted in a defined way and transmitted at one of a limited set of telecommunications carrier and subcarrier frequencies. Downlink telemetry must also be handled in formats, frequencies and data rates compatible with existing station designs, or the project must pay for desired modifications.

New missions added to the US Air Force inventory to be controlled from the Satellite Control Facility in Sunnyvale, California must adhere to a number of restrictions imposed by the existing set of computers, software, plans and procedures that govern all the other spacecraft flown from that facility. In fact most organizations involved in multi-satellite control will have such restrictions. In many cases, existing operations software, which performs such functions as orbit determination and command sequence generation, are passed along from one mission to the next. Although they may undergo some modification for the new mission, their usage restricts the new mission to many of the same interfaces as the old one. Therefore, when the MOS design engineer performs a functional analysis in designing a new MOS, the limitations imposed by the existing physical environment will influence products such as those in Figures 2.5 and 2.6.

2.8 Operations plans and procedures

When the design of the MOS is complete and implementation is underway, the operations system designers can begin to break the operations concept down into one or more operations plans as illustrated in Figure 2.4. The set of operations plans describe the activities of the flight team to accomplish spacecraft and experiment operation. The individual plans will describe the intended organization and structure of the groups within the team, roles and responsibilities of various positions within each group, staffing levels envisioned for each element, and high-level activities to be performed by the group or organizational element covered by the plan. These activity descriptions should contain the following information:

- Activity Description Summary. A paragraph or two describing the objective or purpose of the activity and what functions are performed. Sometimes an activity flow diagram is provided to help in understanding the activity.
- Inputs. A list of all the input required to perform the activity. This includes computer files generated by the performance of prerequisite activities, as well as manually generated input originating either from hand computation or extraction from written reports.

- Outputs. A list of all output produced by the activity. This can include computer-generated files for transmission to other activities, completed forms containing information generated by the activity provided to another activity as an input, printed or written reports, and hand or electronically generated data plots. An output may also be verbal, as in the giving of permission to perform another activity such as the uploading of commands to accomplish an event on the spacecraft.
- Process. The steps within the activity to process the input and produce the output. These may include executing computer software, performing manual computational functions, conducting meetings, completing forms, writing reports or any other action that must be accomplished to satisfy the objectives of the activity. Process description is the heart of the operations plan in that it identifies, in paragraph form, the steps the team members must take to accomplish each task.
- Frequency. A statement as to how often the activity must be performed.
- Performance requirements. An identification of any specific requirements, such as timing or accuracy, that govern the execution of the task. Tasks that must be completed within a certain time limit or prior to other activities will be noted.

A sample activity flow diagram is given in Figure 2.8 for an activity associated with determining whether or not a spacecraft's solar panels need to be offset in their angle to the Sun, implementing such an offset, and monitoring the spacecraft telemetry to ensure that it is accomplished. Activity flows show how the input is operated upon by the process steps to produce the output. They use symbols similar to software flow charts to indicate the type of product or task, and to distinguish between manual and software activities, between computer files and hand-produced data, and between data output to screen, electronic file or hardcopy printout. They establish the logical flow of tasks and show divisions in the flow where operational decisions must be made. For complicated activities, flow diagrams are an essential part of the process leading to the accurate and efficient production of detailed task performance procedures.

After operations plans are complete, the set of operations procedures are written to implement the activities described by the plans, as illustrated in Figure 2.4. Procedures are very detailed step-by-step instructions showing how the activity is to be performed. They contain a checklist so that each step is physically checked off by the person performing the task as the step is accomplished. Every time the activity is performed, a new copy of the checklist is used. Sometimes, in the case of critical procedures, a second person is also required to initial the steps.

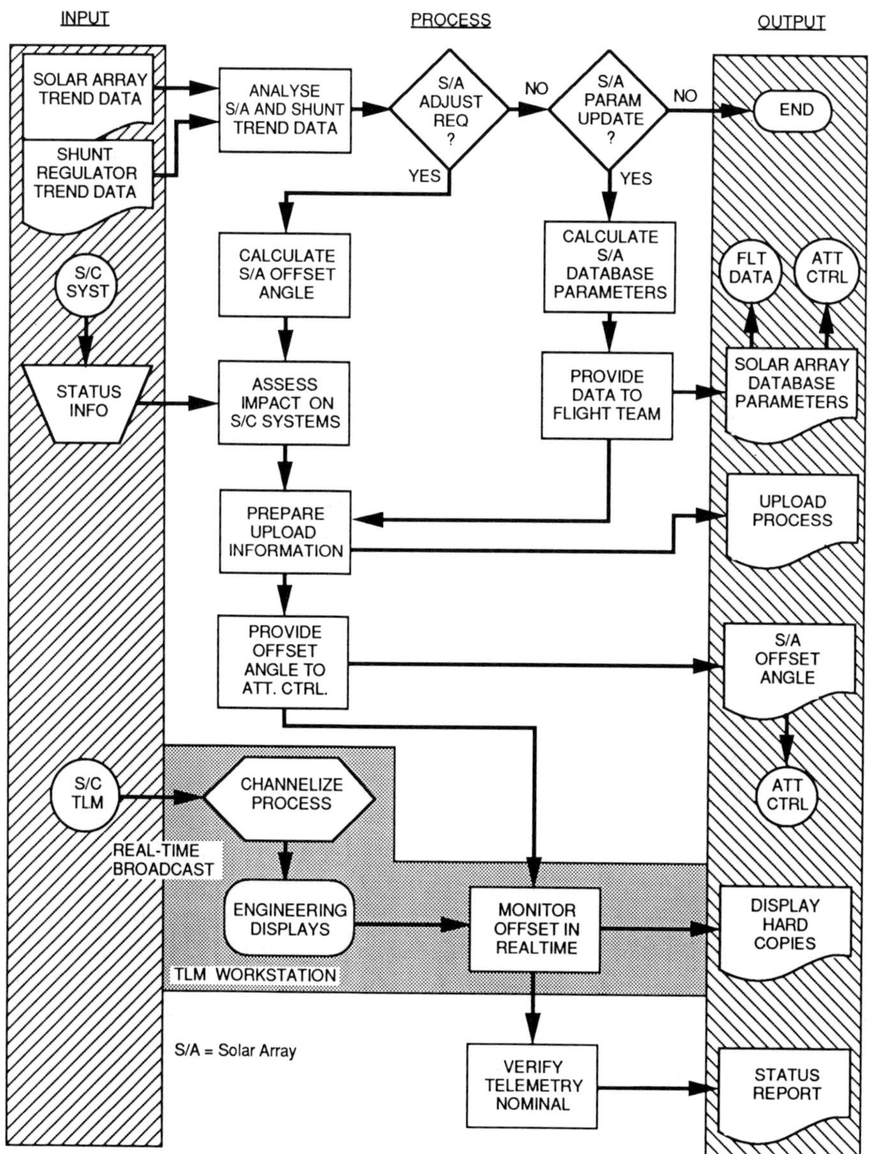

Figure 2.8 Solar array offset activity flow diagram.

Procedures will contain a front section describing the scope and objective of the procedure, and identifying the participants who are involved in executing of the procedure, the required software and hardware, any applicable documents or mission constraints, any prerequisite activities that must be accomplished before this procedure can begin, and any precautions about which the procedure executor must know before he begins the tasks. This is followed by a detailed list of all external inputs necessary to perform the procedure, including file names and identifiers (if known in advance) and report names and numbers. The procedure steps are then given in time order and logical progression so that each step clearly specifies one and only one action. If options are involved that make a difference in how subsequent steps of the procedure are performed, the option selected must be clearly noted in the procedure copy as the selection is made. Key parameter values important to subsequent steps are also indicated in spaces created in the procedure specifically for this purpose. Figure 2.9 is a sample page from a typical operations procedure from the Magellan programme. This particular procedure is one for performing propulsive spacecraft manoeuvres.

2.9 Timelines and scenario development

Timelining and development of operations scenarios are two tools that are used to validate the design of a mission operations system. Timelines are diagrams that illustrate the logical, time ordered sequence for accomplishing the key activities. One axis is time and the other contains activity bars for each task. Each activity is placed in chronological order, showing relational dependencies, so that if task B is dependant on the completion of task A, the former will not be shown as beginning until the latter is finished. Time lining provides two significant benefits for an operations system design. First, it validates the product flow by creating a visual representation of the task durations and dependencies. If a step is missing or chronologically misplaced, it is easier to identify on a timeline than in a mere list of tasks. Second, it provides an indication of potential bottlenecks in the overall process, where several products must emerge at the same time in order for the next step to proceed. Time lines illustrating the two principal components of operations, uplink and downlink, will be further discussed in Chapters 4 and 5.

Scenarios are 'what-if' examples of usage of a mission operations system, where a specific mission is postulated and details are developed against the operations concept, the design and the plan, to see if any flaws in the design logic can be uncovered. These test scenarios can delve into much greater detail than could an attempt to define all possible combinations of missions and activities that could be imagined. By playing one (or several) detailed 'what-if-it-were-to-happen-this-way' scenarios against the MOS design,

OPERATING PROCEDURE
MGN-SFOP-8-120

Step 015 Receive baseline delta-V magnitude from Initial Manoeuvre Profile File (NAV-07)

Enter Delta-V magnitude (m/s): ____________

____ INTL

Step 016 Enter predicted spacecraft conditions and baseline manoeuvre requirements into the Manoeuvre Performance Input-File (TCMx=Pxx.INP) using VMFRIEND SOFTWARE.

____ INTL

Step 017 Execute VMPROP using the Manoeuvre Performance Input-File and the Magellan Feedsystem-File. Review output for reasonableness.

____ INTL

Step 018 If necessary, revise the Manoeuvre Performance Input-File, rerun VMPROP and review output.

____ INTL

Step 019 Prepare the following information for the Delta-V Manoeuvre Data File, Performance Format (SES-101) and deliver to Systems Engineering.

ITEM	TYPE (DIM)	UNITS	DESCRIPTION	CHECK
FMAG	R(25)	Newtons	Effective thrust magnitude.	____
FLORAT	R(25)	kg/sec	Effective propellant flowrate	____

____ INTL

Step 020 Present Performance File (SES-101) data at manoeuvre planning meeting #1.

____ INTL

Figure 2.9 Sample operational procedure page.

potential problem areas can be uncovered and fixed before they are encountered in reality. Many times these scenarios are constructed from prior missions or combinations of prior missions of which the scenario builder has intimate knowledge. For example, an MOS to operate a lander on the surface of Saturn's moon Titan (a follow-on to the Cassini mission) could be tested with a scenario constructed from a combination of the Surveyor moon lander and the Viking Mars lander programmes.

2.10 Summary

In this chapter, we have defined the standard process for design and development of a mission operations system, describing the programme phases, required reviews, documentation and tools necessary to achieve a complete, efficient and cost-effective design. We have emphasized the importance of an early and continuing operations concept, and stressed the need for complete definition of, and agreement on, the functional requirements before the design has proceeded too far to easily modify. We have described how to perform functional analysis, interface definition and analysis, plan and procedure development, time line generation and scenario development to assist in the MOS design, development and validation process.

Table 2.3 is a list of the key terms and concepts presented in this chapter. In the next chapter, we will turn to the people involved, and discuss how to build a flight team to use the MOS we have designed.

2.11 Exercises

(1) From the flow of programme phases and reviews illustrated in Figure 2.2, discuss how they might differ as applied to provision of a science instrument for inclusion on an Earth resources satellite versus implementation of a ground data system for the same satellite.

(2) As the end user of a contractor-developed, turn-key ground data system to operate an ultraviolet spectrometer on board an Earth resources satellite, you are asked to perform the user acceptance testing. What would you put in the user acceptance test plan and report at the acceptance test review?

(3) As the implementation of a mission operations system is nearing completion, what benefits derive from having kept the operations concept document up to date as an accurate representation of the way the MOS will be used to conduct the mission?

(4) Explain why it is important to obtain complete specification of requirements before design begins. Several types of requirements are discussed in this chapter. Is any one type more critical than the others? Justify your answer.

(5) Select one of the level 1 columns (for example, '3.0 Manage Experiment Operations') in Figure 2.5 and expand it into the next level of functional decomposition. The resulting diagram will be one level deeper in the func-

Table 2.3 Key terms from Chapter 2

A-, B-, and C- specifications	*N*-squared chart
acceptance test review	operability
activity	operational requirements
concept review	operations concept
configuration control	operations readiness review
data flow diagram	operations plan
design review	operations procedure
design specification	output
document tree	performance requirements
function	physical functional model
functional analysis	process
functional hierarchy	requirements review
functional requirements	scenario
interfaces	system development
interface analysis	system integration
interface requirements	system operations
input	system planning
logical functional model	system test
mission operations	timeline
mission requirements	voice network

tional hierarchy than Figure 2.5, with the X.0 function occupying the top box, the X.Y functions horizontally underneath, and functions X.Y.1 through X.Y.*n* below those.

(6) Select one of the level 1 columns (for example, '3.0 Manage Experiment Operations') in Figure 2.5 and create a data flow diagram showing the information that must pass between its constituent functions. The resulting diagram will be one level deeper in the functional hierarchy than the data flow diagram in Figure 2.6.

(7) For the logical functional model presented in Figure 2.5 and the resulting data flow diagram in Figure 2.6, design an *N*-squared chart showing the principal interfaces between the five first level (X.0) functions. (Hint: Your completed diagram should resemble Figure 3.8 in Chapter 3.)

(8) Explain how the design of a new stand-alone mission operations system would be influenced by the requirement to use the existing set of eight Air Force remote tracking stations.

(9) Using the single page from the Magellan operations procedure given in Figure 2.9, design a partial activity flow diagram for this activity.

(10) Select any six connected boxes in the activity flow diagram in Figure 2.8 and write as many steps of the operations procedure as you can, to represent this portion of the diagram.

References

Hyman, R. and Ledbetter, K., 1984a, *Command and Control Operations Concept Study, Functional Analysis*, Computer Technology Associates, Inc., GSFC Contract NAS5-27684, Task 500-03a, (March 1984), Greenbelt, MD: National Aeronautics and Space Administration.

Ledbetter, K., 1984b, *Command and Control Operations Concept Study, Operations Concept Methodology*; Computer Technology Associates, Inc., GSFC Contract NAS5-27684, Task 500-03b (July 1984), Greenbelt, MD: National Aeronautics and Space Administration.

3

Organization, Management and Staffing

This chapter discusses the various organizational structures that have been implemented for spacecraft mission operations systems, and the significance of such structures to operations. We initially investigate the basic organizational philosophies derived from functional analysis, defining key management and operational positions and their shifting strategies, for a typical mission control centre. We next examine organizations of various remote-sensing spacecraft missions for differences, and for strong and weak points. Discussed are spacecraft operations conducted by GSFC which controls most US Earth-orbiting science satellites; JPL, which controls US planetary missions; the United States Air Force (USAF), which controls military satellites from the Space Command; and, to a limited extent, commercial communications satellite operations.

3.1 Functional divisions of activity

Organizational structures of mission operations teams for remote-sensing missions are usually based on two top-level (or level zero) functions that are the central focus of any mission of the type under consideration:

(1) To keep the spacecraft (platform) safe and functioning; and
(2) To collect, process and analyse the remotely sensed data.

Both of these functions require the implementation of two distinctly different processes, dividing the system based on the type and direction of data flow (Haynes, 1985): *uplink*, the definition, preparation and transmission of instructions and data to the spacecraft; and *downlink*, the collection, transmission and processing of data from the spacecraft.

Both level zero functions have uplink and downlink elements, as indicated by the matrix in Figure 3.1, and they can be subdivided in several ways which will dictate the project's specific organizational structure. By performing a functional decomposition, as described in Chapter 2, the level zero functions can be broken down into lower levels, as follows. To keep the spacecraft safe and functioning at all times there must be established a 24-hour continuous ground control function where downlink information is obtained, monitored and processed to permit maintaining the spacecraft's

	SPACECRAFT OPERATION AND SAFETY	COLLECT AND PROCESS SENSED DATA
UPLINK PROCESS	1. COMMANDS FOR ENGINEERING OPERATION OF SPACECRAFT 2. MANOEUVRE COMMANDS 3. TELECOMMUNICATIONS COMMANDS 4. EMERGENCY OR ANOMALY RESOLUTION COMMANDS 5. ENGINEERING DATA LOADS (EPHEMERIS, MEMORY MGMT, STAR REFERENCES, ETC)	1. COMMAND SEQUENCES FOR EXPERIMENT OPERATION 2. ON-BOARD DATA STORAGE COMMANDS 3. PAYLOAD POINTING COMMANDS 4. LONG-TERM PLANNING FOR SCIENCE OPPORTUNITIES
DOWNLINK PROCESS	1. MONITORING OF HEALTH AND STATUS TELEMETRY FROM SPACECRAFT SUBSYSTEMS 2. SUBSYSTEM TREND ANALYSIS 3. SUBSYSTEM PERFORMANCE PREDICTION 4. QUALITY OF DOWNLINK SIGNAL	1. EXPERIMENT DATA COLLECTION, STORAGE, AND TRANSMISSION 2. DATA PROCESSING AND ENHANCEMENT 3. IMAGE PROCESSING (IF APPLICABLE) 4. DATA QUALITY ASSESSMENT

Figure 3.1 Relation of data flow processes to top-level functions.

health and safety via subsequent uplinked instructions. Therefore the first level zero function can be expressed as the following two level one functions:

(1) Establish mission control operations for spacecraft contact.
(2) Maintain spacecraft health and safety.

For scientific missions, either Earth-orbiting or planetary, expedience is not usually required in the processing and analysis of the sensor data, the second level zero function, unless the mission is adaptive (see Chapter 6). However, in order to identify which data to collect and how they can be obtained and brought to the ground, planning and scheduling functions are necessary to provide coordinated input from the off-line science activity into the real-time operations of the spacecraft. Therefore, to the above are added these additional level one functions:

(3) Plan and schedule mission data collection activities.
(4) Process and analyse received sensor data.

Lastly, there must be a coordination, direction and control function that unites these four segments of mission operations into a cohesive, efficient unit. This is the function of mission management and it becomes the last level one function:

(5) Manage mission operations activities.

Figure 3.2 illustrates the placement of the five level-one functions into a functional distribution for typical remote-sensing missions. Under *mission management* are four divisions, each defined by a set of level two functions shown in the figure, which were developed by another level of functional decomposition and which indicate the content of each area. Note that the association of the activity on the ground with spacecraft activity in flight

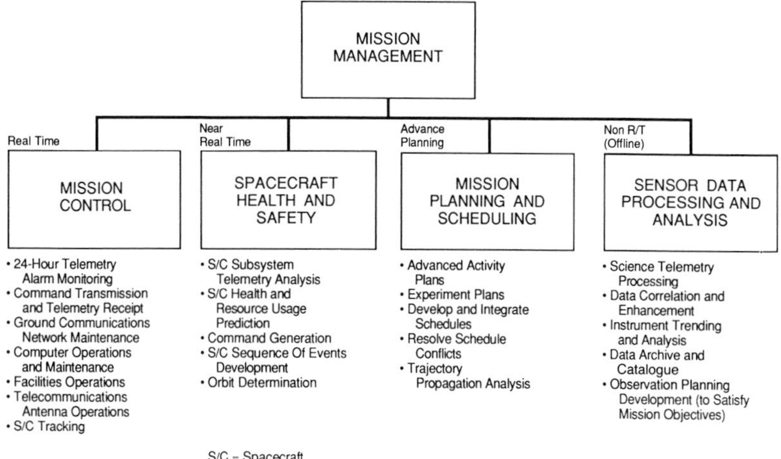

Figure 3.2 Typical functional distribution for a scientific remote-sensing mission.

decreases in timeliness from left to right. *Mission control* is charged with real-time control of the spacecraft, including commanding and real-time monitoring of its telemetry, and the day-to-day operations of the ground control centre. No attempt is made in this function either to contemplate why actions are taken or to perform any analysis of downlinked telemetry — this function serves only to operate the spacecraft as would the driver of a car. However, first indications of platform health problems are detected here and alerts given to the spacecraft technical support teams.

Spacecraft health and safety is responsible for in-depth monitoring, analysis and maintenance of flight hardware and software, for prediction of future performance, for determining the current trajectory ephemeris, and for building the command files to be uplinked. This function operates in near real-time, monitoring live telemetry when necessary and interfacing with mission control for command files to transmit to the vehicle. *Mission planning and scheduling* operates a cycle ahead of the activities occurring in flight and provides the framework within which real-time and near real-time operations work. This includes both long-term planning activities to decide how best to conduct the mission for optimal science return, as well as near-term weekly operations schedule generation to keep operations activities functioning. *Sensor data processing and analysis* operates furthest from real time. Here the user's downlinked bits are processed into science data products and interpreted for mission results. Here also are housed the user's representation to other project elements.

There are several functions included in the operations concept discussed in the previous chapter which are not covered by the functional distribution in Figure 3.2. These include ground system development (which is often still occurring or under upgrade even as the spacecraft is flying), personnel training and evaluation, logistics (which includes supply, transportation and handling), and facility maintenance. In general, satellite control centres are located on existing facilities, either NASA or military, and such generic services as maintenance and logistics are provided by the base organization. For those areas that may be tailored specifically for a given programme, such as ground system development or training, a staffed function to mission management is usually shown.

A significant factor in determining how to convert the above functions into an organizational structure is whether or not the mission is conceived of as *adaptive*. In an adaptive mission, the processed information from one data collection period is used in some way to decide how to collect data during a subsequent period. Commonly, adaptive missions may use the contents of images to re-target subsequent images or other investigations. As an example, a primary goal of the Viking lander investigation was to search for the existence of life on Mars. Thus the operations system was organized to allow the landers' cameras to be quickly repointed at anything which might be identified as a possible life form in the initial images. Complex sequences which directed the surface sampling arm to collect samples were begun based on what was seen in the images. Weather satellites and military reconnaissance satellites may have similar adaptive requirements.

Adaptivity is distinguished from the need to respond quickly to spacecraft or other anomalies discovered in the downlinked engineering data. This latter need is likely to exist in any mission, and the organizational structure must be able to respond to such situations independently of how it responds to science data collected. Details of adaptivity and anomaly response mechanisms will be discussed in Chapter 6.

For many scientific missions, either Earth-orbiting or planetary, adaptivity is either not required or expressly forbidden in order to simplify the mission operation. Magellan, for example, would be capable of using its SAR data to improve navigation accuracy, or of using its altimeter data to improve SAR commanding. To avoid adding mission complexity, thereby saving cost, neither of these capabilities was implemented. The rationale is that expedience is not required in the processing and analysis of the data collected, and there is no requirement for a quick turn-around interface between data analysis and mission planning.

3.2 Organization from functional analysis

Given the functional distribution in Figure 3.2 as representative of most remote-sensing missions, how do we go about converting it into an organization that is functionally efficient? Comparisons to historical mission organizations show that some functions, such as 'spacecraft health and safety' nearly always convert one-for-one from functional to organizational entity. Many of these can be readily recognized in our simplified mission organiza-

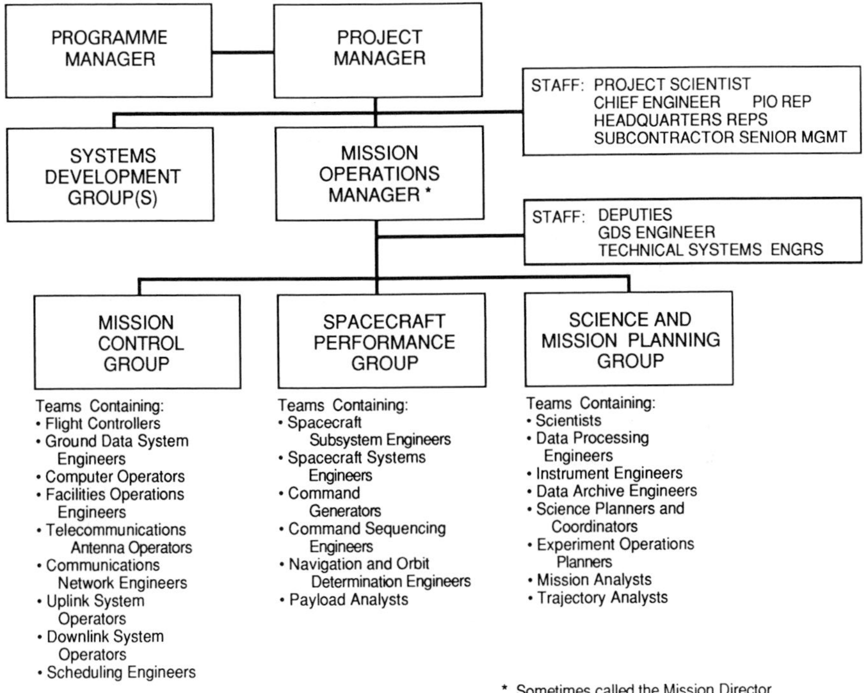

Figure 3.3 Simplified organization of a typical scientific remote-sensing mission.

tion resulting from this conversion process, as illustrated in Figure 3.3. 'Mission control' also largely translates across. Others, such as 'sensor data processing and analysis' are harder to trace directly, although for scientific remote sensing, one can generally assume that 'sensor data' implies science.

Past missions have shown a general consistency in organizational structure. However, through the history of mission operations organizations, four tasks have not been consistently placed in the structure, and some discussion is necessary.

First, navigation, the function of determining precisely where the vehicle is located as a function of time, has sometimes been elevated to a top-level activity. However, the function lends itself to distribution, as shown in Figure 3.2. Collecting spacecraft tracking data is performed in real-time by the personnel that operate the telecommunications antennas. Orbit determination is a process that ingests the tracking data and grinds out, usually via an off-line mainframe computer, the current ephemeris, very similar to the processing of subsystem telemetry in near real-time to determine the status of a subsystem. Then, the mission planning group must look at the propagated trajectory to determine how the experiments must be scheduled to accommodate where the vehicle is going. We have chosen to reflect the navigation function split among these three groups in our typical organization of Figure 3.3.

Second, the 'mission planning and scheduling' function is probably the most volatile, having been placed at various levels and often even split apart. Mission planning has occupied a position on a level with mission control, spacecraft health, and sensor data analysis, as Figure 3.2 indicates; has been subordinated within one of the other three; and has once been located as staff to mission management. Using the experience of many past missions, we have separated mission planning from scheduling and placed it where the majority of projects have located it, with science data analysis in a Science and Mission Planning Group (see Figure 3.3) linked closely with the user, because in our experience this juxtaposition has served well the purpose of letting the user guide the mission plan. Scheduling activities for daily operations has been relocated to the Mission Control Group where it can quickly react to the volatile nature of operations.

Third, designers of detailed vehicle sequences-of-events have been located organizationally in both mission control and spacecraft performance functions. This task is generally performed more efficiently placed together with the spacecraft analysts, as we have shown it, but in adaptive missions it can be more nearly a real-time task and perhaps is more properly put with other real-time activities of mission control. Finally, data management (including archiving) has been placed under either mission control or with the science group. This choice largely depends on how closely that task is associated with real-time operations. Specific examples will be discussed later in this chapter.

With these considerations, and based on the functional distribution in Figure 3.2, the chart in Figure 3.3 represents the organization that most nearly reflects the typical scientific remote-sensing mission. Sections 3.3 and 3.4 will discuss this organization in detail.

3.3 The management element

The management structure of a mission operations programme can make or break the success of the mission. Organization, control and authority begin at the top and the perception of these to the working troops in the operational trenches can have a large effect on how well their job is performed. Proper attention to pyramid organizational structure, clear lines of authority and communication, and other principles of good management, important in any organization, are particularly vital when decisions must be made correctly and quickly. Operations personnel tend to be highly dedicated, highly motivated individuals who must be listened to by management as well as firmly directed if they are to function well.

At the top of the organization of almost any mission lies its sponsor, which may be governmental (as in the case of NASA or the military) or may be private industry. In either case, the sponsor will generally appoint a Programme Manager who has charge of policies and funding, and, in the case of a scientific mission, a Programme Scientist. In general, the Programme Manager does not in any way get involved in performing the mission. Depending on the size of the effort, the sponsor may also have staff such as financial analysts.

In general, the management element of the performing agency consists of a Project Manager (PM), a Mission Operations Manager (MOM), called the Mission Director in some control centres, and several managers at the next level, which have been called directorates, offices or groups. The PM has the ultimate responsibility for the spacecraft and the data return to the sponsoring organization. With such responsibility comes the absolute control over who is a member of the flight team and what is done with the spacecraft or the instruments.

The PM is the mission's representative not only to the sponsor but also to the outside world; he is the spokesman for every event that occurs, both good and bad. He tracks both the events and the expenditures incurred in getting the events accomplished. To do this, he must understand the total programme at the functional level and be continually informed through the use of status summaries. Major decisions that affect mission objectives, spacecraft health or modifications to the data collected will be made by the PM based upon recommendations from the MOM and other members of his staff.

For very small science projects, the Project Scientist will usually manage the project. With sufficient funding, however, one of the first divisions of labour that should be considered is to separate these two duties, with the Project Scientist on the staff of the Project Manager. In this way the science goals of the mission will be tempered by the rules of good management. The Project Scientist is the representative to the science community for the project and its scientific goals, and serves as a filter of both science objectives to be achieved and the scientists who will staff the flight team. For larger projects, the Project Scientist may have on his staff an advisory group called a Project Science Group, consisting of some members of the sponsor-chosen science team.

Also important to the balance of objectives is to have a Chief (or System) Engineer staffed to the PM, to keep the interests of the flight vehicle in balance with those of the mission's science goals. In the past, large projects have involved several separate contracting companies — perhaps one or more who constructed the payloads, and another who built the carrier platform. Since there must be a single manager who carries ultimate responsibility for mission success, senior management from all principal contracting companies will be on staff to the Project Manager. This permits rapid response from a contractor if a problem arises with a component they have built or personnel they have provided to operations. The staff may also contain the Public Information Officer for communicating mission progress, mission problems, and scientific results to the press, particularly in NASA missions, and possibly, but less likely, for military missions. If required, representatives from the mission funding agency headquarters may also be on the Project Manager's staff.

A systems development organization is also shown reporting to the Project Manager in Figure 3.3. If the mission operations system is developed incrementally or experiences nearly continual upgrades during a lengthy lifetime, this organization takes the responsibility of accomplishing this development and smoothly integrating the upgrade into operations. This organization reports to the Project Manager rather than to the Mission Operations Manager, because it usually occurs off-line to normal day-to-day operations. Obviously development activities take lower priority than active vehicle operations unless the product under development is critical to an upcoming in-flight event. This function includes design and development of both hardware and software and may involve modifications to the operations concept or other documentation such as the operations plans and procedures.

While the Project Manager is usually involved in interfaces external to the project, day-to-day control of flight team operations is performed by the Mission Operations Manager. The emphasis of the MOM is on successful performance of the mission as defined. He makes all decisions on commands to be sent to the spacecraft to control either vehicle performance or data collection. He supports the PM as necessary, but in general is not concerned with external interfaces. The MOM's staff usually has one or more deputies, due to the potential for the necessary presence of a MOM at any hour of the day or night. Also on his staff will be the technical systems engineers for the spacecraft and components of the ground system, such as the telemetry processing system or the telecommunications antennas. Some of the principal scientists may also act in an advisory capacity. Separation of operations from management by one level not only allows one of the two positions to keep the overall goals in sight while leaving the other free to deal in crisis management, but keeps science (or other user) and operations equivocal in high-level decisions.

3.4 Operational personnel positions

Directly below the Mission Operations Manager (or Mission Director) in the organization in Figure 3.3 are the three basic functional elements containing the groupings of personnel that perform the mission. The chart is generalized in that for a large project each group will consist of several teams of people dedicated to a portion of the task assigned to the group, whereas for a smaller project one individual may take on several positions. A sample of the specific personnel positions within each group is listed under the group, although the actual number of individuals will vary widely with the size of the project.

The Mission Control Group contains the *flight controllers*, the 24-hour-per-day personnel who watch spacecraft subsystems and sensor telemetry for any potential problems. The authority of the flight controller to take specific action varies widely between missions, all the way from personally making an in-flight correction to being limited to notifying the proper persons to make the correction. Sometimes spacecraft or instrument engineers supply this group with contingency response commands or command loads for immediate transmission if a certain anomalous condition occurs. In our typical organization, this group also contains schedulers who provide operations with the coordinated activity schedules necessary to conduct efficient, error-free operations.

Other around-the-clock personnel positions are those that staff the various components of the ground system that must continue to operate without interruption. Examples of these are the telemetry, tracking and command (TT and C) operators, ground communications network engineers, computer operators, antenna operators, and certain facility operations engineers. Usually, the only flight team positions required to be staffed 24 hours a day, seven days a week, would be those in the Mission Control Group. However, depending on the mission or mission phase, multiple shifts might also be required of some groups or positions in other groups.

Positions in the Spacecraft Performance Group may or may not be required to operate around-the-clock, but it is assumed that whenever the spacecraft is performing a significant action, such as a manoeuvre, members of the spacecraft team will be present. The heart of this group is the personnel representing each of the spacecraft's subsystems, typically expert engineers in power, propulsion, thermal control, telecommunications, attitude control, data handling, flight software, and mechanisms, depending on the characteristics of the particular spacecraft under control. Also included are systems engineers who look at each planned action from the perspective of the overall spacecraft rather than one specific subsystem; sequencing engineers who design and build the command files to be transmitted to the vehicle; and navigation or orbit determination analysts who maintain accurate knowledge of the position of the vehicle in space and plan for trajectory correction manoeuvres as required.

The Science and Mission Planning Group is responsible for taking the decommutated, decalibrated sensor data from the telemetry processing subsystem on the ground and converting them into data products for analysis and interpretation. This group may contain data processing technicians, image processing and enhancement engineers, instrument engineers, and science observation planners. It will contain the teams of university scientists associated with the mission experiments who conduct the scientific analysis and interpretation of the data. Larger projects will include in this group representatives for the science teams who act as interfaces to the remainder of the project to interpret operations activities to the scientists (who probably have not participated in operations training exercises) and to represent the scientists in activity decisions at operational meetings. Also in this group, for reasons previously stated, is the mission planning function, containing the planners who provide the critical link between the scientists who interpret the mission objectives and the engineers who must implement it. This group also contains the mission and trajectory analysts who provide time histories and predicts of spacecraft geometrical relationships with respect to other solar system bodies for mission planning use.

3.5 Personnel shifting strategies

All spacecraft mission operations centres must have the ability to monitor and, when necessary, contact the vehicle at any hour of the day or night. Most are required by the nature of the spacecraft trajectory to have regular contacts scheduled at hours other than the standard work day. No matter how hard the operations crew tries to keep all significant contact activities confined to normal working hours, occasionally, if not more often, a critical manoeuvre must occur outside of normal working hours. Therefore, most MOS designs include planned 24-hr positions and must design shift strategies.

Of course this concept is not new or confined to spacecraft operations. Many plants, major airports, police departments, and telephone operators are just a few of the occupations that require around-the-clock work shifts. Most of these use the standard day shift, swing shift and graveyard shift based on local time, and for many applications in spacecraft control, these are also acceptable. The flight controllers normally utilize the standard shifts. However, there are some applications where deviation from this strategy better serves the mission.

As an example, several of the experiments aboard the Viking landers on the surface of Mars functioned either better or exclusively during Martian daylight, where the lander activity schedule peaked. The length of the solar day on Mars is 24 hr, 39.6 min, close enough to Earth's day that the daily high activity period was delayed by only 40 min per Earth day. Wanting to concentrate ground efforts on this high activity period, mission designers initially defined a 'prime' shift that started 40 min later each day. Although the landed mission was begun this way, human nature soon rebelled at having to reset alarm clocks daily, and people's internal biological clocks were continually confused. At a rate of 40 min per day, in 12 days the high activity daylight period moved later by one complete 8-hr shift, so mission planners redefined the 'prime' shift to one that only rotated every 12 days. The personnel monitoring lander activity could now change shifts less frequently. At the other extreme, the Shuttle Imaging Radar (SIR), a NASA programme to operate a series of state-of-the-art synthetic aperture radar sensors using the Shuttle as a platform, operates its instrument for less than 10 days. In such cases, two 12-hr shifts are practical, thus providing the advantages of reduced workforce and a smaller number of shift handovers.

The typical JPL mission allows the numerous teams to set up their own shifting schedules, as long as the necessary personnel are on station for all activity required by their charter. Consequently, there are always a variety of shifting strategies to cover 24-hrs and seven days. In attempts to share the pain of non-standard shifts among everyone, eight-hour shifts that start at noon, 8 pm and 4 am are evident as are ones only shifted 2-hrs from the more traditional starting times. To cover weekends and still give all personnel an occasional weekend off, seven days at work and three days off were common as were ten days working and four days off. A rule of thumb for covering three 8-hr shifts lasting for more than a few weeks is that five individuals are required to be available for each position so that days off, sickness, holidays, and other absences can be accommodated.

3.6 The large, dedicated programme

There are numerous examples of large, dedicated programmes: GSFC programmes such as Hubble Space Telescope; JPL programmes such as Voyager and Galileo; and military programmes such as Navstar global positioning system (GPS). The characteristics of these kinds of programmes that make them large are as varied as their mission objectives. Viking had to control four planetary spacecraft, two of which were on the surface of Mars and subject to variable weather conditions. Voyager had only two spacecraft, but encountered four planets and tens of planetoid-sized moons from distances where the one-way travel time of the communication signal was measured in hours. When completed, the final GPS configuration will contain at least 18 Earth-orbiting satellites broadcasting navigational information, controlled by three dedicated ground up link stations and eight tracking monitor stations.

We have selected the Viking project, a programme of NASA's Langley Research Center, as representative of this class of mission operations systems. Figure 3.4 summarizes the organization that was responsible for the two orbiters and two landers (Robins, 1976). Viking included over 800 people on its flight team, to support the vehicles with their 10 lander science experiments and 4 orbiter science experiments. Twenty per cent of the team was in the mission control group, 37 per cent performed the spacecraft health and safety function, and 38 per cent constituted the science and mission planning group. The remaining 5 per cent were staff positions for mission management, including a contingent that ran a simulator at the lander contractor's facility.

Despite the sheer size of the organization, it functioned quite well. It symmetrically consisted of three directorates, each containing three groups. Each group contained as few as two teams with a maximum of twelve in the Lander science group. The orbiter and lander systems analysis teams were by far the largest, both subdivided into units according to vehicle subsystem, each of which had the size and influence of other teams. The Mission Control Directorate included any activity that related to real-time operations and a

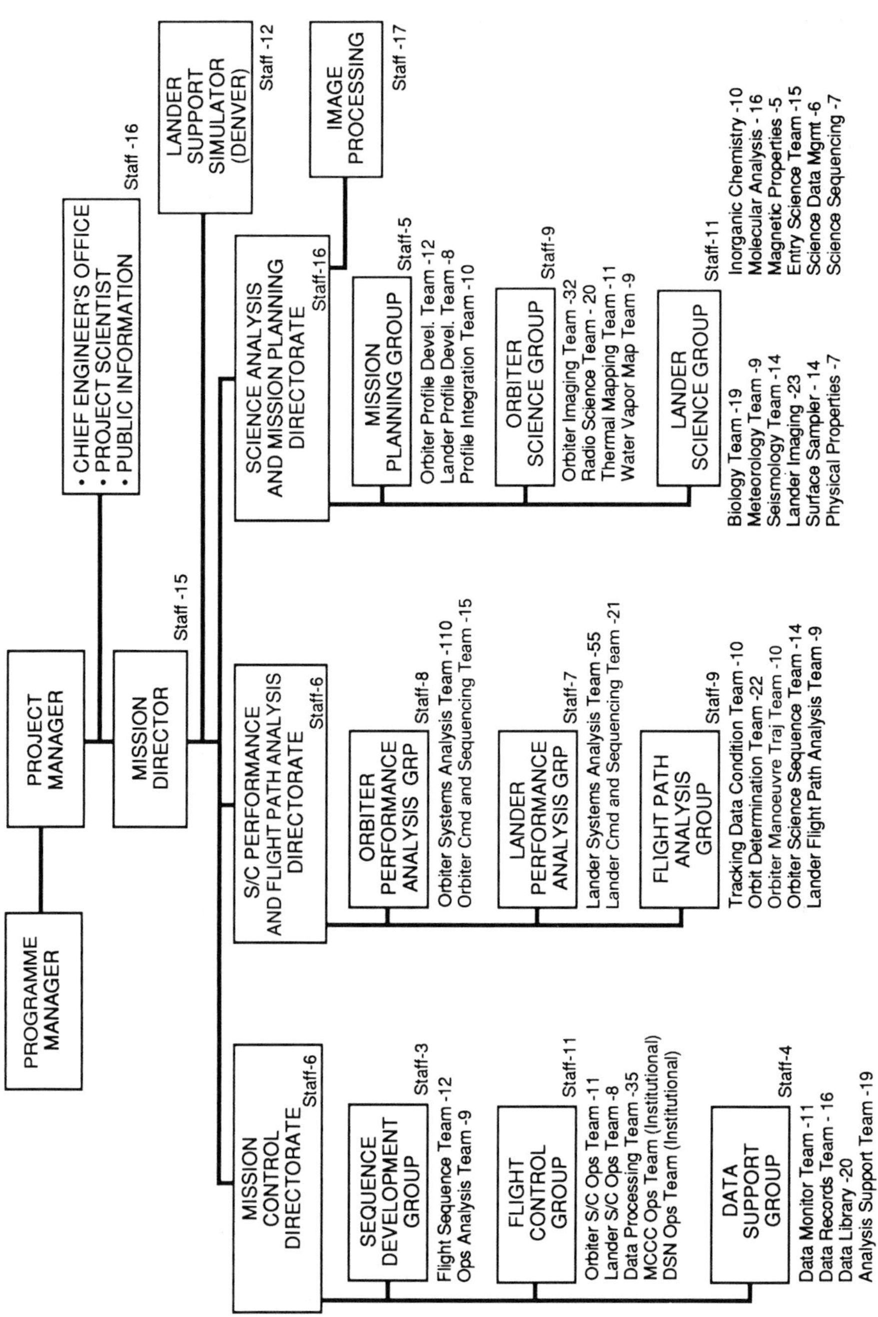

Figure 3.4 Organization of the Viking Mars mission.

few other peripheral activities. The Spacecraft Performance and Flight Path Analysis Directorate (SPFPAD) contained the analysts for both vehicle subsystems' health and their positional accuracy. The Science Analysis and Mission Planning Directorate contained all the science activities from planning upcoming data collection activities to analysing the final processed data.

There are a few differences between this organization and the ideal functional distribution of Figure 3.2. As was the case in our typical organization, the generic mission planning and scheduling function was split and recombined with other functions: mission planning was included with science analysis and scheduling was performed by the operations analysis team in the sequence development group of Mission Control. Having planning and scheduling in separate directorates created some interface problems that might not have existed had they been combined.

The data archive and catalogue (or library) function was under Mission Control in the Viking scheme rather than science data processing, probably because it handled all the magnetic tapes used by the ground processing system, which operated 24 hours a day. In spite of this, the major activity in the data support group was in off-line support of the science teams. In an idealized system, this function properly belongs with them.

The navigation-related functions of vehicle tracking, orbit determination and flight path analysis were gathered into the flight path analysis group of SPFPAD rather than distributed according to timeliness of operations.

Since Viking was the first attempted landing on another planet, with the eyes of the world on the flight team, there was redundant staffing for some functions. For instance, there were usually at least four different teams that monitored telemetry in real time during important spacecraft events on one of the vehicles. In the case of the orbiter, the orbiter spacecraft operations team and the orbiter performance analysis group were charged with real-time monitoring of the health and status of the orbiter subsystems, and a corresponding team and group monitored lander activities. The flight control group's chief mission controller or 'ace' and the data monitoring team also monitored activities for both platforms, although the emphasis of the latter team was on data quality and its progress through the ground system rather than on spacecraft health.

3.7 *A small, economical programme*

Although there are many large, well-known programmes, there probably are a greater number of small programmes that have been just as successful. For such missions, costs must be carefully controlled, and the operations system must be kept small. However, if the principles of this chapter are applied, operations can be carried out economically. One such program is the SIR project, whose flights to date have been operated from a payload operations control centre (POCC) located at NASA's Johnson Space Center. For SIR,

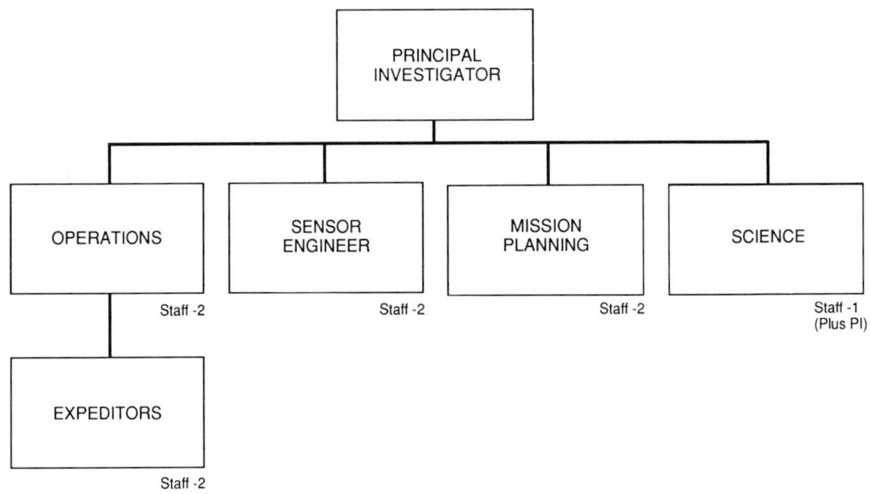

Figure 3.5 Organization of the SIR-A mission.

the platform is operated by the Shuttle operation team, and the only operational responsibility for the SIR flight team is to the sensor and associated systems.

From an organization standpoint, SIR provides an interesting example because the flights of SIR-A and SIR-B, occurring in 1981 and 1984 respectively, and of SIR-C, planned for 1994, have increased steadily in both complexity and available resources. SIR-A operations system organizational structure is shown in Figure 3.5. The Principal Investigator served as Project Manager and one member of the two-person science group. Each of the other groups consisted of two members as well. Each member worked one 14-hr shift, providing one hour's shift handover twice a day. The resulting activity was intense but sustainable for the short periods of time. The operations group contained the SIR flight controller, who performed the command functions and served as chief interface with Shuttle controllers. A sensor engineering group monitored sensor health and validity using samples of the science data which were returned in real time, and verified correct response to commands issued. The mission planning group used a small minicomputer to predict Shuttle position with time and performed the functions of orbit propagation, activity plan generation and modification, schedule integration and sequence design. Primary data from the sensor were collected on board using an optical film recorder and returned to the team after landing. A data librarian collected both film and ancillary information (orbit history, command records, etc.) after the fact. Two expeditors brought the total flight operations team count to ten individuals. The experiment was operated successfully except for minor Shuttle fuel cell problems, which forced termination of the mission after 3 days.

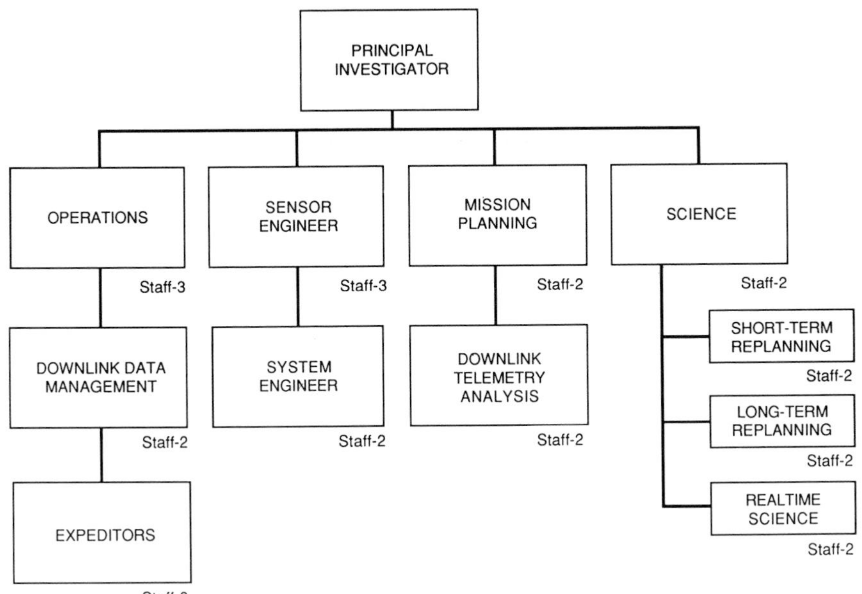

Figure 3.6 Organization of the SIR-B mission.

The SIR-B flight experiment increased the complexity of the sensor by adding a pointing system and a digital data system, including an on-board tape recorder and a data downlink system. A science team of 46 experimenters was added, and worldwide field experiments were coordinated with *datatakes* (data-gathering events) to provide ground truth. The operations team size was more than doubled; the resulting organization chart is shown in Figure 3.6. A systems engineering position was added to perform off-line analysis of downlink telemetry. The operations, engineering, and mission planning positions were changed to 8-hr shifts as a result of the SIR-A experience. Two additional mission planning positions were added so that the activity plan could be modified in near real time. Another added position called 'Real-time Science' was added to direct field experiment coordination, and two positions were created to monitor engineering and science data recording on the ground.

The SIR-A experience showed that it is the nature of Shuttle flights to require changes to operational schedules, and therefore several unconventional steps were taken with the SIR-B organization in order to facilitate quick re-planning. The mission-planning function was charged with schedule integration and orbit propagation as before, but re-planning was done under the science function to allow direct interfacing with field experiments. Field experiments are large and costly operations in their own right, and the re-scheduling of a single datatake for SIR often translated into major personnel, equipment, and even ship movements for those in the field. Thus re-planning had to be done with great care and constant communication with field locations. On another subject, rapid re-planning was also required in response to downlink telemetry when, for example, minute differences between orbit predicts and the actual orbit required real-time commanding to keep the sensor recording meaningful data. This loop had to be kept very tight so the downlink telemetry analysis task was allocated to the mission-planning function.

The SIR-B flight lasted 7 days, and although several problems were encountered in both platform and payload which forced complete replanning of the mission in near real-time, the unconventional design of the operations team permitted the mission to be carried out successfully and most of the originally-planned data were collected.

At this writing, SIR-C is planned to follow in 1994, with repeat flights of the same sensor beyond 1995. Again, the SIR-C sensor has been improved, with the addition of multiple wavelength and polarization channels and corresponding increases in downlink and on-board recording capabilities. Modifications to the flight team structure to reflect these additional changes in the sensor are anticipated.

3.8 The military way

The United States Air Force controls the vast majority of military spacecraft put into orbit by the US under direction of the USAF Space Command which has its headquarters in El Segundo, California. Although the USAF has satellite control centres throughout the world, the majority of their spacecraft are controlled from either the Satellite Control Facility in Sunnyvale, California (a single six-storey, medium-blue building with no windows, universally called the 'Blue Cube') or the newly developed satellite control facilities at Peterson Air Force Base in Colorado Springs, Colorado.

There are several significant differences between the organizations of military spacecraft operational control centres and those in the NASA world. First, the managers are all military officers who may or may not have been involved directly in spacecraft operations earlier in their career. Second, the turnover rate of operating personnel is generally higher than for non-military centres. Third, the objective of most military missions is to obtain and process remotely-sensed data very quickly, so that in general there is higher pressure on downlink schedules.

As with NASA missions, military control centres are usually located on existing military bases where logistics support and facility maintenance are taken care of by base operations. This permits prime concentration on the operations of the mission itself. Figure 3.7 presents a simplified organization for a dedicated Air Force spacecraft control operation. The Control Centre Commander, similar to the Project Manager in civilian operations, is usually a full Colonel. He is responsible for both the support activities and the satellite control operations. He delegates the actual satellite control operations

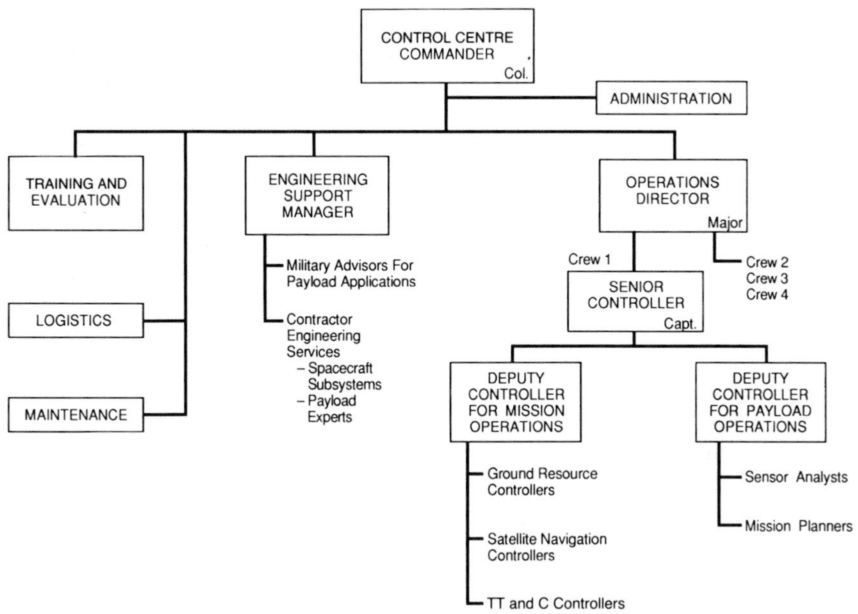

Figure 3.7 Simplified organization of a military remote-sensing mission.

to an operations director. In most cases, the Control Centre Commander did not advance from an operational position and therefore has not had the experience necessary to conduct day-to-day operations. He focuses his attention on the administration of control operations and ensures that the proper engineering support is provided.

The Operations Director, similar to the Mission Operations Manager, and usually of the rank of Major, is more likely to have had prior operations experience. Reporting to him, in most cases, are four rotating crews of controllers who actually perform daily satellite operations. Each crew is headed by a Senior Controller (a Captain), who monitors the activities of both mission operations and payload operations.

Mission operations, in military applications, are defined as those necessary to keep the spacecraft and ground system performing so that the mission can continue. For most military control centres the contingent of personnel necessary to perform this is limited to the Ground Resource Controller, the Satellite Navigation Controller and the TT and C Controller. The engineers who are the spacecraft subsystems and sensor experts are not a part of this operations organization, but are separated under engineering support. This is because they are largely made up of civilian contractor employees from companies that designed and manufactured vehicle components who are under contract to provide support services. They are segregated from the military controllers largely for security reasons. This is a major difference

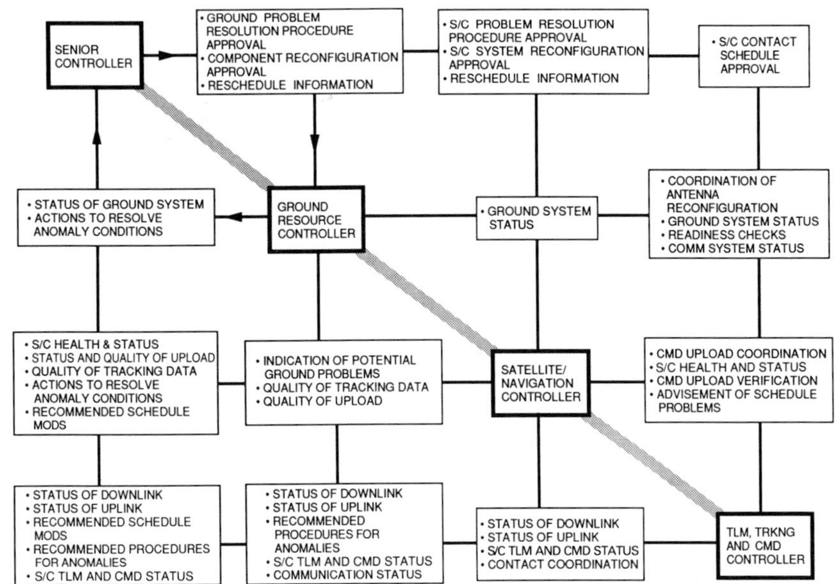

Figure 3.8 Control room interaction during routine operations of a military mission.

between NASA and military satellite operations. Figure 3.8 is an *N*-squared chart that identifies some of the interfaces between the real-time operational positions in a typical military control room.

Payload operations comprises the analyses of the data from the on-board sensor and the operation of the sensor itself. The people who perform these tasks are the military equivalent of the NASA scientists who interpret the returned information and decide what to do next with the mission. They may also have advisors and other support personnel in the engineering support organization. This type of organization is conducive to adaptive missions where sensor results can be quickly fed back into the planning process to change the spacecraft's upcoming activities. This is somewhat similar to the change mentioned in Section 3.7 for the SIR-B organization where the planners who were needed to achieve rapid rescheduling were a part of the science group.

3.9 Commercial operations

The past 20 years have seen an explosive growth of the commercially operated communications satellite, with several companies following the lead of Comsat Corporation into the business, and with many countries owning geosynchronous communications satellites. The next 20 years may see a similar growth in other types of Earth-orbiting satellites owned and operated

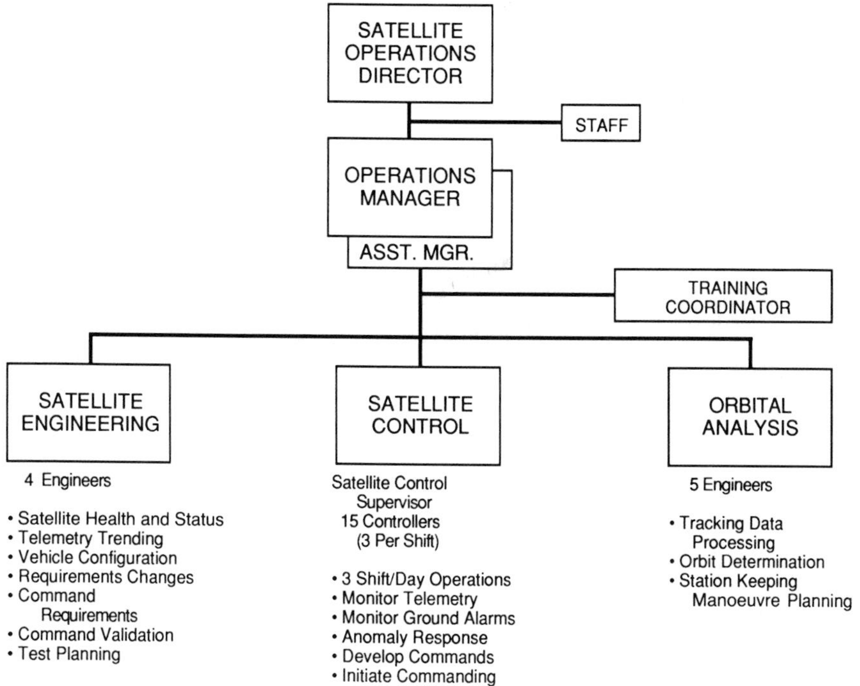

Figure 3.9 Typical organization of a commercial communications satellite operation.

by private enterprise. When profits form the motivation for accomplishing a task, a company can get very creative in organizing for operational efficiency.

As an example, Hughes Communications, Incorporated operates up to 13 communications satellites out of their control centre in El Segundo, California. They have been able to combine functions, cross-train operations personnel, and install sufficient multi-satellite operations to minimize the number of required personnel, thereby minimizing the costs necessary to operate the satellites and their mission. Their communications satellites are of three different series, operating over at least three different bands of the frequency spectrum. All are spin-stabilized and operate at geosynchronous altitude. The organization and staffing for this control centre is in Figure 3.9, showing the division of activity into three areas: satellite engineering, satellite control, and orbital analysis. Satellite engineering is responsible for maintaining the configuration and health of the spacecraft and its subsystems. Satellite control is the 24-hr a day monitoring and control of operations. Orbital analysis tracks and maintains the orbit of the satellites to within the user's specifications.

Note the relatively small number of operations personnel (approximately

28) for controlling 13 spacecraft. There are several reasons for this economy of operation other than the fact it is a commercial activity. First, the spacecraft are all similar and have a single purpose, as opposed to a one-of-a-kind planetary mission. Second, there is no data analysis segment of the organization. Since communications satellites merely pass along data to the ultimate end user, no data processing and analysis personnel are necessary. They occasionally must support troubleshooting of data problems if the user does not receive the data in the condition he expects, but this is for anomalies only. Also, there is a minimal mission planning effort as compared with a NASA science mission. The vast majority of mission planning occurs before the satellite is launched, by engineers who complete their job and do not remain in place for flight operations. Once the satellite is inserted into its slot in a geosynchronous orbit, the mission planning effort is over unless the customer for the vehicle wants it moved to another location later in its life. Last, and perhaps most important, is that the operations personnel are all cross-trained to respond to any of the satellites the control centre is handling, and the satellite engineers are therefore knowledgeable about all subsystems on each spacecraft. Although the Hughes training program is such that a new person can be functional and make a contribution in as early as 4 months, the full training programme requires about 13 months for all satellites.

3.10 The ideal operations organization

Like any other organization, operations organizations must function efficiently and cost-effectively. Thus, the same principles of good organizational design that apply anywhere are important here as well. Good organization benefits operations systems especially in the following areas:

1. A defined organization agreed upon early in the mission design phase simplifies the writing of all operations documentation because it delineates lines of authority and establishes interfaces that need to be defined as operations nears. It also allows required attendance at meetings to be limited to those who need to inform, those who need to be informed and those who make the decisions. Corollary to this point is that the organizational structure should be general enough that one structure can survive as the project transitions through design, development and operations phases of a mission operations system. For example, a development-oriented organization can begin with a position reporting to the MOM during system development and smoothly evolve into an operations orientation with a development position on staff to the PM, as shown in Figure 3.3, as the MOM becomes involved with actual operations.

2. In times of anomalies requiring quick action, a clear understanding of the lines of authority (who directs who) makes communication and the flow of direction well understood so that they need not be argued in real-time. This does not mean that a good manager should not listen to his subordin-

ates; it does mean that subordinates must understand who makes decisions (and which ones are made at each level of the organization) after all the arguments have been heard.

3. Clear delegation of responsibility allows upper levels of management to be saved from resolving minor conflicts, leaving them free to manage the major ones. A Project Manager cannot afford to manage the project alone (unless the project is very small indeed). A good organization chart will force him to delegate authority.

Items 2 and 3 can be largely satisfied by maintaining good pyramid shape to the organization. This means that the number of positions reporting to a superior position should not vary widely from top to bottom of the chart, and every position should have at least two subordinate positions. There will, of course, be exceptions to this rule.

The second point that has become clear is that an organizational chart must never be confused with a data path chart. Just as telephones may connect every position with every other one, data flow and even decision flow will occur between organizationally separated positions. But subordination cannot allow this. Each position on a project can only serve one master. One exception is that funding source and line-management functions (such as hiring, termination, promotion and raise review) need not follow the project organization. Several major aerospace companies operate well using 'matrix' management structure which intentionally has project management separated from line management to provide a check and balance on task definition and resource requirements.

As previously discussed, many projects operate using individuals who are employed by different subcontractors mixed into the project organization. Viking functioned extremely well with NASA personnel subordinated to Martin Marietta and JPL managers, even though Martin Marietta and JPL were contractors to NASA. The Project Manager insisted that the organizational structure be designed to avoid company allegiance interfering in the mission operations. Key positions were filled with the best people available, regardless of their institutional home.

Finally, a few specific lessons which experience has taught should be mentioned. In both short-duration projects like SIR and periods of intense activity surrounding key events in longer term projects, there is sometimes a tendency toward what might best be described as hysterical behaviour which must be avoided at any cost. It is during times of crisis where mission-ending critical mistakes are made. This tendency is probably not unique to remote sensing missions. The best way to conduct 'hysteria management' lies in a well-designed organizational structure which allows a sense of order to be maintained and through which crises can be managed. A separate but related subject is that the imposition of organizational and schedule structure can be important enough to require its own organizational element. After launch, the Magellan project created a team called 'mission engineering' whose purpose was to develop and track scheduling and timelining activities using

Table 3.1 Key terms from Chapter 3

adaptive mission	Programme Manager
Chief Engineer	Project Manager
Chief Scientist	Public Information Office
cross training	scheduling
data processing and analysis	science analysis
downlink process	sensor engineering
flight controller	shifting strategy
matrix organization	spacecraft health and safety
Mission Control	uplink process
mission management	
Mission Operations Manager	
mission planning	

traditional techniques such as integrated flow diagrams. Perhaps the biggest lesson of experience (and one that is difficult to quantify) is that any step which can be taken within the organization to create a sense of unified purpose is probably worth taking. Engineering tutorials to scientists, science tutorials to engineers, participation in selected reviews and decisions by all project personnel, and similar activities have resulted in cross-fertilization and information sharing that have contributed to each member feeling more a part of the whole, and have materially benefitted several projects.

Operations organization has often been a highly-debated topic within projects. During its design phase, Magellan changed organizational structure not less than three times, and with dozens of proposed organizations, before deciding on a final structure. Good organization produces better work, higher efficiency, and more productive, happier personnel. Although it is true that a bad organization can always be overcome if the personnel are sufficiently motivated and capable, it is always at the expense of time and money, which are the two most valuable consumables a project has.

3.11 Summary

This chapter has discussed organizational structures of mission operations systems for large and small NASA remote-sensing missions as well as military and commercial satellite operations, contrasting the styles and functionality of each. We have examined the benefits and disadvantages of various ways to structure a flight team and pointed out the key roles and responsibilities of the major positions and groups. Lessons learned from past experiences have been expressed as cautions for operational organizations of the future. Table 3.1 is a list of the key terms and concepts presented in this chapter. Next, we will turn our attention to how the organization achieves command loads to the spacecraft and payload to perform the mission, as we describe the uplink process.

3.12 Exercises

(1) Flight team organizations are usually built around the functional divisions of activity (columns in Figure 3.1) rather than the processes. Could an organization be constructed about the processes? If so, what are some of the difficulties such an organization would experience? If not, explain why not.

(2) The Navy is planning to launch a satellite to monitor hurricanes in the Atlantic. The spacecraft contains an imaging system that must be used by ground operators to search for tropical storms and, upon finding one, send commands to image the various parts of the storm with high resolution visible-light and infrared images. The mission must therefore be adaptive. How would you modify the organization in Figure 3.3 to reflect this mission? Redraw the chart and give justification for each change you make.

(3) Saturn's large moon Titan has a solid surface and an atmosphere over twice as thick as Earth's. NASA's Cassini mission will put an orbiter around Saturn and drop a probe to Titan's surface. If a Cassini follow-on mission is postulated with a lander using an imaging system that operates only in the presence of sunlight, what weekly and daily shifting strategies would you recommend for the flight team to implement activities and monitor telemetered data, given that Titan's solar day is 16 Earth days long?

(4) Given the directorate divisions for the Viking project indicated in Figure 3.4, design an *N*-squared chart similar to Figure 3.8 that identifies the major operational interfaces between the Viking Mission Director, the Mission Control Director, the Spacecraft Performance and Flight Path Analysis Director and the Science Analysis and Mission Planning Director.

(5) Give some reasons why the location of the mission planning and scheduling functions are so volatile when trying to fit them into an organization.

(6) Comparing the organizations and operations of the Viking and SIR projects, what suggestions could you provide that might reduce the size, complexity and cost of a large project like Viking?

References

Robins, C. H. Jr., 1976, *Viking 75 Project, Viking Flight Team Organization and Staffing*, M75-150-5, Viking Project Office, NASA Langley Research Center (23 June, 1976) Hampton, VA: National Aeronautics and Space Administration.

Haynes, N. R., 1985, Planetary mission operations: an overview, *Journal of the British Interplanetary Society*, **38**, 435–438.

4

The Uplink Process

As discussed in Chapter 1, the focus of a remote-sensing mission is to acquire remotely sensed data from a payload instrument on board a platform. Both directing the payload for data acquisition and operation of the platform require the two basic processes, uplink and downlink, that were introduced in Chapter 3. This chapter discusses the uplink process. The next will address downlink.

The uplink process is the portion of the operations system which flows from the user's need for an on-board activity through the planning, creation and transmission to the platform of the set of commands which instruct platform and payload to meet the user requirement by accomplishing the activity. The goal of the process is to accomplish the activity within flight hardware constraints and resource limitations in accordance with mission safety requirements. The uplink process has an urgency, or at least a potential for urgency, associated with the entire process which can give it a very different flavour than downlink. No matter how routine the nominal mission goals may be, spacecraft operations require that the entire uplink operation always has the capacity for reaction to problems.

In the early days of remote-sensing missions, payloads were relatively unsophisticated. As a result, the mechanism required to drive these missions was not stressed, and sophistication was not a high priority in the uplink process or in the platform computer systems. However, as missions have developed over several decades, spacecraft payloads have evolved into very flexible and programmable devices. Concurrently, demands for reliability on both ground and platform portions of the uplink process have increased. More stringent requirements for error-free commanding, flexibility in on-board execution of commands, automated checking for both invalid commands and for adherence to overall mission rules regarding safety and longevity, and cost-efficient payload operations have served to increase the complexity of uplink systems. The limiting factor in mission accomplishment has become the lack of speed and versatility in the uplink process, whether in command preparation on the ground or within the platform data systems. From an MOS standpoint, the technical challenge in future uplink process design is to make sure that the process is designed and operated so that the limiting factor for the mission is the level of sophistication which can be designed into the payload, rather than the speed or capacity of any component of the ground or platform support system.

As with all elements of operations, desire for cost effectiveness has put pressures on uplink systems to operate not only well but efficiently. As we progress into the age of commercial exploitation of space, this pressure will no doubt increase. To this end both NASA and its contractors have studied and implemented several cost-reducing concepts in the past 10 years. Most of these fall into four categories:

(1) increasing the inheritance in both flight and ground elements from mission to mission,

(2) developing multi-mission systems which not only serve several missions at the same time but also become permanent facilities which decrease the need for mission-unique system development,

(3) developing distributed mission operations systems which decrease travel and relocation costs, and

(4) incorporating new command philosophies involving expert systems or knowledge-based tools.

Some of these are discussed in more detail in Chapter 8.

The past and present missions we discuss in this book have had widely diverse objectives. The uplink systems of each have been highly tailored to meet their individual requirements, and the resulting structures were quite different. However, even though some of the operations system designs were done independently and with no common root, they shared a basic structure. This chapter describes those functions common to them all. It is important to note that timescales for each step, number of occurrences of each step, staffing levels, degree of automation and many other parameters have varied widely with the implementation for each mission, from the 6-day Shuttle flights to planetary orbital and landed missions, some of which have been operated for more than a decade.

Of particular effect on the uplink process design is whether the mission is required to respond to its own activities — in system design terms, whether it operates as an open or closed-loop process. The three primary occasions which call for closed-loop operation — anomaly response, adaptivity and field experiment coordination — are mentioned here only briefly and are given a more thorough treatment in Chapter 6.

Another important decision, to be made by smaller projects, is whether the timing of command execution is to be done from the ground or on board the platform, or whether both modes of operation are required. Large projects generally have done both. It is always advisable to be able to send commands for immediate execution whether or not they can also be sent for execution at some later time because of the need to respond to emergencies. If telecommunications systems can allow constant contact with the platform and if precise timing of commands or sequences is not required, the mission operations system may be able to assume the responsibility for direct command issuance. If both of these conditions do not hold, a simple command sequencer may suffice. If repetitive sets of commands are required, for example to perform an instrument start-up sequence, they can be stored on board and

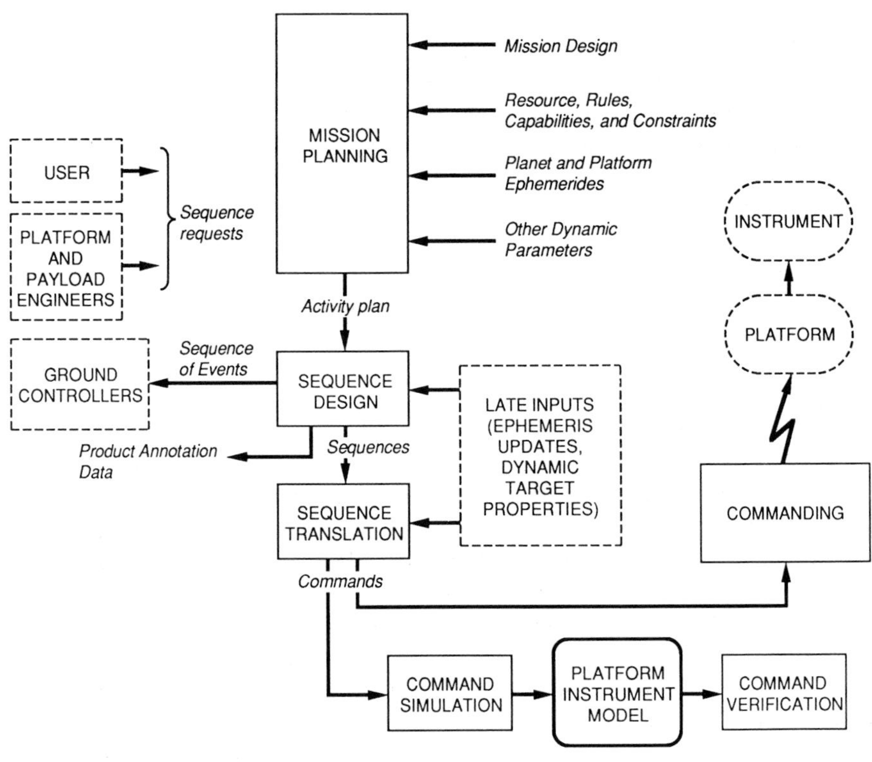

Figure 4.1 Data flow in the uplink process.

sequenced with a single command using the same sequencer. Forethought in both platform and payload design can greatly simplify mission operations.

A generalized data flow chart for the uplink process is shown in Figure 4.1. Elements which are associated with the flow but are not part of it are shown in dashed boxes, and products which flow from one step to the next are shown in italics. Broadly, requests are provided to the *mission planning* task by two groups: the science teams or other users of the remote-sensor data, and the engineers who are responsible for the operation and maintenance of platform or instrument. The function of the mission planning task is to balance these requests against: (1) the realities of geometry, such as target and platform positions, viewing periods, planetary occultations, etc.; (2) capabilities of the platform, payload(s), telecommunications link, MOS and other elements (including limitations due to the use of expendables on either platform or payload); and (3) constraints imposed by project policies with regard to resource conservation or safety.

The goal of the mission planning task is to create an *activity plan* (sometimes called an *advanced sequence plan*) which is consistent with the constraints and resource limitations of the platform and sensors and which satisfies user requests to the maximum extent possible. After careful review for completeness, the activity plan is passed to the *sequence design* task. There, with encapsulated specific knowledge of platform and instrument design and operation, command sequences are generated which will perform the requested activities. The *sequence translation* task then prepares the bit patterns and bit error codes for the actual commands to be uploaded and executed, adding late inputs only as necessary. Before these commands are transmitted to the platform, ideally they are executed in a *command simulator* — a model of the flight system which will show how the actual platform will respond to them. Only after verification that the requested action is correctly produced by the simulator is the command load sent to the platform or instrument(s) for execution.

The process outlined in the figure can represent any time scale from hours to months, depending on the requirements of the project. Thus the process flow could represent any point on a spectrum from the planning of an entire mission, the planning of one specific mission phase such as an encounter, or a single command load. The process can also be required to cycle in ways not shown in the figure whenever an output of a step is shown to contain errors or inconsistencies: a failure in command verification, for example, will force reruns of whatever processes are required to resolve the failure.

4.1 Mission planning

The first task in the uplink process is to plan the mission phase under consideration. General mission objectives established for NASA projects by a science steering group, and more specific strategies defined by the NASA-selected project science group (also called science working group or investigator working group) are used as a starting point. Specific data-taking requirements are outlined by individual investigators or discipline groups. Major objectives are traded against principal resources, resulting in a prioritized list of goals which can be used later in lower level trade-offs. This necessarily involves interaction between the end user and the mission planner, who must be aware of platform and payload capabilities and limitations, mission restrictions, and the consumable resources of the entire system. Since there are almost always more requirements for activity on the system than resources to satisfy them, mission planning generally involves multiple iterations among user, planner, and platform requirements to resolve conflicts.

The inputs to the mission planning task are:

(1) the mission design;

(2) knowledge of the spacecraft and object-of-interest geometry pertinent to the phase;

(3) a description of policies and constraints, including allowed and prohibited mission activities (preferably written in a document agreed to by all elements of the project);

(4) a preliminary list of requirements placed on that portion of the mission being planned by both users of the payload and operators of payload and platform. This latter list will include major mission events (such as launch), sensor and platform requirements, and user objectives, all at whatever level of detail is available.

The goal of the task is to understand the requirements and to identify those which, to first order, can be met within the policies and constraints.

The mission planning task can be broken down into four steps: collection of requirements, identification of opportunities, definition of resources needed to satisfy requirements, and resolution of conflicts at the activity plan level. These steps are seldom executed serially, and the strength of each is highly dependent on the specific nature of the mission. The first step generally proceeds with a series of meetings with users. An established science team provides a detailed science plan for the phase, often with the aid of science workshops. Early involvement with the project mission planning team in these discussions keeps focus on those objectives which are realistic given the capabilities of the mission. The current state of knowledge about the body or field under investigation is summarized, and objectives are formulated.

The Voyager project held such workshops in 1975, 1978, 1984, 1986 and 1989, before each planetary encounter. These were followed by a series of Planetary Discipline Working Group meetings, where compiled lists of specific objectives were generated. Major conflicts among working group desires can be negotiated at this level, and a written statement of objectives can be created for use by mission planners. In the Voyager case a preliminary sequence of observations was created at this point. Similar techniques have served as starting points for both planetary and terrestrial missions (e.g., Miner *et al.*, 1985; SIR-B Science Team, 1984). Early written summaries of likely characteristics (brightness or reflectance ranges, sea states, etc.) of the targets to be sensed can be easily assembled during these preliminary discussions and later used in the uplink process to set instrument parameters such as gains. Unrelated to the uplink process, such summaries can also benefit instrument designs.

Following definition of science requests, engineering requirements must be integrated into the developing activity plan. Platform and payload engineers are considered users of uplink resources too, and must formulate their own plans to equivalent detail. In some cases where the platform and payload share command systems, platform requirements (and crew requirements if they exist) must also be integrated at this point. In other situations such as instruments to be flown in the Shuttle, platform operators have a separate allocation for their uplink requirements, and these are integrated into the command flow later in the activity plan development. This step produces a preliminary activity plan that can then be merged with datataking opportunities.

As a part of this integration, it is common to introduce personnel with both scientific and instrument expertise to add ancillary requirements such as calibration, instrument characterization and health checks, and the like. These individuals, often called Experiment Representatives or Science Coordinators, must be cognizant with science objectives and payload capabilities so that they can represent both in later negotiations with project personnel, ensuring that science goals are not compromised as the more practical concerns for safety and equipment health are integrated.

The second step of the mission planning task is the identification of opportunities for satisfying user requests. For some mission types (e.g., mapping missions), objectives may be general enough that they are not a detailed function of time, and this task is not a difficult one. Landed missions and some orbital missions are also unconstrained enough that periods of time can be freely allocated to user objectives. However, in general there are some constraints placed on the activity plan by geometry. Available solar power, encounter approach geometry, planetary or solar occultations, and other similar constraints can dictate observation times. Definition of times for these observations generally requires software predictors which determine, and often graphically illustrate, the observation geometry.

A useful aid to this task is an *opportunity timeline* (Linick, 1985), a tool to identify or define opportunities for taking data. Opportunity timelines are produced by software that generates plots of encounter geometries, including predicted occultations. More sophisticated tools also identify opportunities given the required datataking geometries such as lighting and viewing angles. These have been particularly useful in the SIR mission, where repeating (or almost repeating) orbits and overconstraining requirements for multiple imaging of targets with varying geometries require such software to optimize orbit parameters. This type of tool, called an *inverse planner*, identifies all possible datataking opportunities given an orbit, a set of targets and a range of permissible viewing geometries (Wilton, 1985; Harris, 1984; see Appendix 1).

To help identify *targetting* opportunities for extraterrestrial missions with the capability to point sensors or groups of sensors, targetting software has been devised to identify potential observation times. Given the platform and target ephemerides, such programs plot potential sensor footprint outlines (image frame, spectrometer track, etc.) on to the surface of a planet or moon and allow the user to manipulate viewing and lighting geometries to define the observation. A program of this type known as POINTER was essential to the Viking orbiter and Voyager flyby projects in identifying the multiplicity of imaging opportunities in the presence of complex geometrical constraints.

After opportunities have been established for both science and engineering requirements, the third step of mission planning, the identification of required resources, begins. In this step, the activities necessary to take advantage of the opportunity are defined and expanded. 'Take colour images of an active volcano on Io' may be a well-defined science goal, for example, but it tells a mission planner very little about the detailed steps and resources required to accomplish the task or whether the activity will create conflict with another requirement to 'image the Jovian ring system in infra-red light'.

Resolution of, or even definition of, conflicts should not be considered until each activity has been defined in terms of the constraining resources required to accomplish it.

Any quantity which has the potential for limiting the capability to fulfill user needs is a constraining resource. Consumables such as fuel or electric power are obviously spacecraft limitations, as are such non-consumables as pointing accuracy and stability. Command and telemetry channel capacities, tape recorder capacities, limited lifetimes of tape recorders and tapes, data-relay link availability and other more subtle effects such as orbit determination accuracy and viewing geometry constraints imposed by bodies such as the Sun, must also be taken into account. Platforms containing multiple remote-sensing instruments must also consider overlapping view space, platform angular momentum, and other resources for which the various instruments will compete. Where the platform is manned, crew requirements must also be defined. Although they do not directly lead to an uplink, ground events (such as allocations of communication link times) must also be roughly coordinated by mission planners in order to discover constraints that are attached to each activity. Both planetary and terrestrial missions may be plagued by a high degree of uncertainty in telemetry link allocations early in the design process, but even rough time allocations should be folded in as early as possible to allow for conflict resolution. It is far easier to adjust detailed schedules later for updated view periods than to fail to identify major resource limitations early while they can still possibly be removed.

A list of constraining resources common to most missions is shown in Table 4.1. These come in several categories. Discrete constraints are those which are applied on an instance-by-instance basis. Two instruments on the same gimbal cannot point in different directions at the same time. A single crew member can only accomplish one function at a time; and a single computer cannot (usually) sequence more than one activity. Cumulative constraints are applied over a period of time, and can be allocated until their cumulative limit is reached. Spacecraft carry a limited amount of fuel; batteries have a finite amount of stored energy before recharging; and instruments have a statistically limited lifetime. Still other resources have *rate constraints* — the spacecraft cannot dissipate more than a limited amount of heat from its electronic and other components, it must limit its power consumption, its turn rates are limited, and so on. Each mission will have its own set of peculiar resource limitations, and it is a challenge for mission planners to identify these as early as possible.

Once it is well defined, each request can be evaluated against established project policies which define acceptable risks, guide use of resources, determine priority of objectives, and in general offer guidelines or directives on how to conduct the mission. The defined requests are gathered into an activity plan and conflicts are identified. The activity plan should be kept in the form of an on-line database so that it can be (1) accessed by software which generates a graphical timeline, and (2) readily changed as it evolves

Table 4.1 Typical constraining resources for platforms and payloads

Discrete

- sequence timing
- crew time
- illumination and viewing geometries
- uplink bits and rate
- downlink bits and rate
- telemetry occultations (e.g., by platform)
- command memory and data storage (e.g., tape recorder) capacity
- platform angular momentum
- view space and direction
- orbit determination accuracy
- ephemeris age
- instrument lifetime
- equipment lifetime (e.g., solar panels)

Cumulative

- fuel
- energy usage
- instrument (scan platform, spacecraft, etc.) lifetime

Rate

- power
- thermal
- slew rates (platform and scan platform)

into final form. Additional software can then total the usage of cumulative limitations and the rates of rate-limited resources, and can identify discrete violations as well.

A full activity plan, containing all known required activities and listing all known resource requirements for each, is essential before beginning the process of conflict negotiation, the fourth step in mission planning. If the database is large, or the mission heavily overconstrained, software must be available to quickly identify constraint violations as well. Without such tools it will be impossible to avoid the unintentional creation of new conflicts when attempting to solve existing ones.

The process of negotiating with users to obtain an activity plan that satisfies objectives within resources is usually iterative and, in the authors' experience, has always been underestimated in terms of required time and personnel. In principle it would be easier to assign each request a priority and to accept them in priority order until the first resource limitation is reached, but in practice this places unrealistic limitations on mission accomplishments. There are far more intelligent solutions. Creative planning sometimes finds ways of satisfying conflicting requests in unforeseen ways. Parameters of the mission, such as orbit period or location of ascending node for orbital missions, can be optimized to meet the maximum number of requirements.

Failing that, conflicts can often be resolved by face-to-face negotiations with users, who can recognize the importance of others' requests and suggest alternative ways of satisfying needs with less or different use of resources. Parameters which had been previously specified as rigid requirements are frequently found to be more flexible when conflicting requests are seen by both sides to have scientific value. Experience has shown that not only is less science lost by such an exercise, but the synergism generated by such discussion often actually increases the net value to both users. Therefore only in extreme cases is it necessary for management to decide between two desired observations. Resolution of conflicts is an important part of the role of Experiment Representatives, especially in closed-loop missions where replanning must often be accomplished in very short time periods.

In the process of building the plan, gross mission parameters such as launch date, transfer orbits, and orbit parameters such as altitude, inclination and ascending node will have been determined if they have not been previously fixed. In simple missions the activities may remain conflict-free, but more frequently it will be necessary to replan activities as the uplink process continues, either due to the discovery of errors or to changing objectives, constraints or other parameters. The primary goal of the activity plan is to provide enough detail to allow negotiations among users and between users and engineers at the major activity level. At regular intervals through the planning process, the activity plan must also be reviewed to verify that user requirements are still accurately represented.

The output from the mission planning task is an activity plan which is free of conflicts at the level to which requests have been developed. This timeline should have been checked for self-consistency and consistency with platform constraints and resource limitations at a rather high level. It should be known to satisfy user requirements and should violate no established mission policies. Ground events should be included in this schedule with whatever accuracy they are known at the time.

An excerpt from an activity plan from the SIR-B mission is shown in Figure 4.2. It will be used as an example throughout this chapter. This excerpt shows four orbits, numbered 92 through 96, during which datatakes (image strips) 92.2 through 96.2 are acquired. Each datatake has its start and stop time defined, with the times of actual data start and stop shown. Instrument look angle and an indication as to whether calibration of data is required are shown, as are Shuttle attitude required to image the target (abbreviated here as 'BLU' or GRN', which means 'payload bay to Earth' or 'wing to Earth', respectively) and the crew wake-up time for this period. This level of detail, and no more, is required for negotiation at this point.

The mission planning task must take concepts and turn them into an executable plan. For some simple projects this task might be strictly a serial one where mission goals are turned into a single activity plan for later translation into commands to the platform and payload. In most cases, however, the activity plan must be segmented into several uploads, if for no

Data take	Orbit	Start data (d/hhmm)	Stop data (d/hhmm)	Attitude	Look Angle	Calibrator
92.2	92	5/1553	5/1605	BLU	50.0	off
92.4	92	5/1615	5/1622	BLU	40.0	off
92.6	92	5/1625	5/1629	BLU	40.0	on
92.8	92	5/1705	5/1721	BLU	31.0	on
93.2	93	5/1732	5/1738	BLU	18.0	off
93.3	93	5/1748	5/1800	BLU	50.0	off
93.4	93	5/1805	5/1810	BLU	50.0	off
93.5	93	5/1817	5/1821	BLU	20.0	on
93.6	93	5/1844	5/1853	BLU	53.0	off
94.2	94	5/1904	5/1910	BLU	25.0	on
94.3	94	5/1920	5/1929	BLU	50.0	off
94.4	94	5/1932	5/1937	BLU	40.0	off
94.6	94	5/1941	5/1946	BLU	57.0	on
95.2	95	5/2036	5/2040	BLU	38.0	on
95.4	95	5/2047	5/2102	BLU	53.0	on
crew wake		5/2100				
95.6	95	5/2111	5/2115	BLU	53.0	on
96.2	96	5/2148	5/2158	GRN	28.0	off

Figure 4.2 Excerpt from the SIR-B activity plan.

other reason than lack of space on board. More often, it is necessary to make changes to the plan as a result of updates to parameters such as platform position or target conditions. A further major complication to serial mission planning is that uploads must sometimes be made on schedules such that the time between uploads is shorter than the time required to complete the mission planning activity.

The solution to this complication is overlapped planning cycles, as shown in Figure 4.3. This figure is a timeline, with time increasing to the right. Progress of the uplink process is shown increasing toward the bottom of the figure. As the figure shows, the mission planning effort has been divided into long- and mid-term activities. The mission plan is generated as a part of the long-term activity. Two mission plans are to be developed, one for each of two mission phases. One phase might represent one planetary encounter, or one major objective such as a landing. The first mission plan is developed (as shown by the dashed line at upper left) and delivered to the mid-term activity so that an activity plan can be created from it.

From that activity plan a total of eight sequences are to be designed, and from each sequence design there are to be three uploads. Each sequence design is based on a new update of the activity plan. After two such updates, generation of a new activity plan is begun. When the new plan is completed it is used as the basis of sequences, with updates to the new plan occurring after each. Preparation of each separate upload occurs whenever dictated by the platform requirements using the latest available plan. In parallel with the above activity, an updated mission plan is developed by long-term planners and delivered to the developers of the third activity plan. While the third activity plan is used for sequence design, development of the phase-two mission plan is underway.

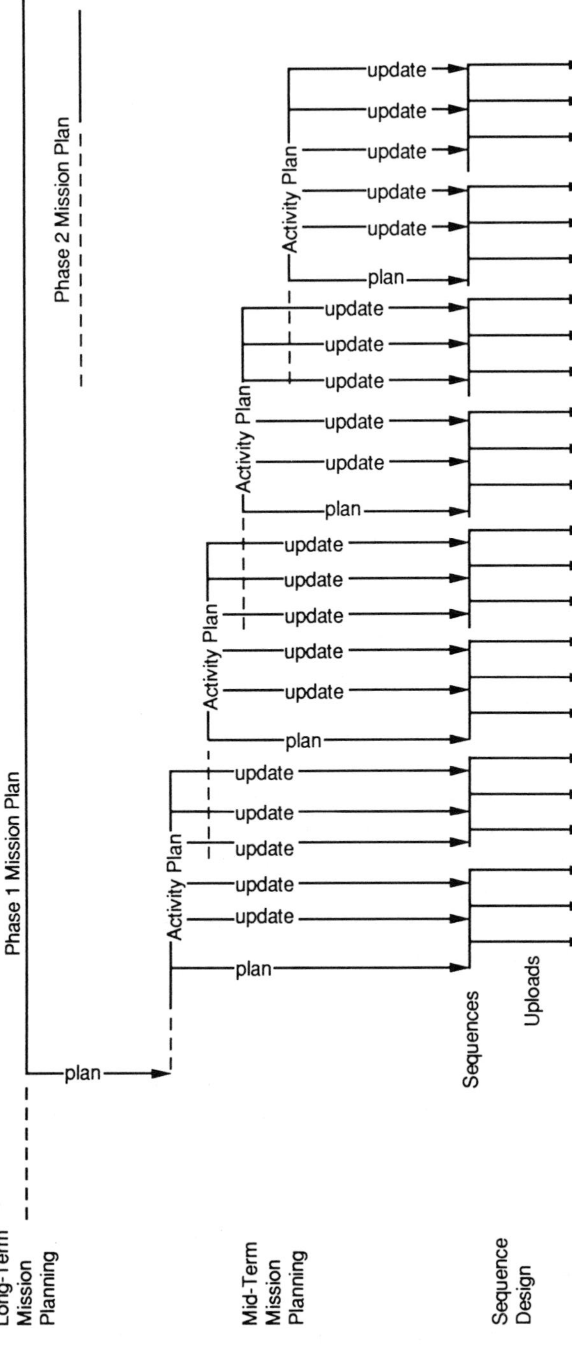

Figure 4.3 *Overlapped mission planning. Development activities are shown by dashed lines.*

Note that in this scheme updates are accommodated all along the way, and that changes in plans can even be accepted just prior to upload generation. Clearly there are limits to what level of changes can be accepted at each level, as each of the steps involved has less and less time to absorb them. The Viking lander's most difficult activity, the acquisition of a surface sample, provides a good example of how these limits change. The activity required (1) several picture-taking sequences to identify an area of the local terrain that might be sampled, and (2) interleaved sequences of moves by the sampling arm with intermediate images which documented the sample before and after acquisition and served to help troubleshoot sequences that failed to get any sample. Mission planning accepted re-plans involving interleaving of surface sampler and imaging activities only at the mission plan level. Such planning required major resource trade offs involving complicated on-board tape recorder management schemes, sunlight illumination angles, and sampler acquisition strategies. Planners would allow simpler changes to sample location during mid-term planning, but they would allow only minor changes such as camera gain settings during the sequence design task.

Design of the mission planning task must take into account the mission's requirement to accept late changes. As we have shown, the flexibility to accept changes must necessarily decrease to zero as the time for the uplink approaches. How quickly that flexibility decreases affects how robust the mission planning and other tasks of the uplink process must be. Figure 4.4 shows schematically how this is true. The vertical axis of the graph depicts (in some arbitrary way) the amount of flexibility allowed in design of an upload. At the top of this axis would lie major replans, and at the bottom minor changes to parameters which do not involve resource trades. On the horizontal axis is time until uplink, decreasing to zero on the right end of the graph. Flexibility as a function of time-to-uplink is shown by two curves for two hypothetical missions. The difference in flexibility between the two missions lies not in the endpoints of the two: they may be equally flexible at inception, and equally fixed at the time of the upload. But in mission (a) resource allocations are fixed early in the planning phase, and the major portion of the work involved in the planning process can be done early. In mission (b) flexibility is maintained until late in the planning task. Thus mission (b) requires more work to be accomplished in a shorter time than (a) (i.e., more work per unit time), and mission (b) is far more expensive to operate. Cost and complexity of the planning task, and therefore of the uplink process, is directly related to the shape of the flexibility — time curve.

When the mission, project and user requirements become stable enough that major changes are no longer occurring at the activity level, the sequence design task can begin. Said another way (and for many time constrained projects, more accurately), when schedules demand that sequence design activities begin, the activity plan should be frozen. Exceptions will undoubtedly occur, where errors are discovered or the associated risk of change is deemed acceptable by project management.

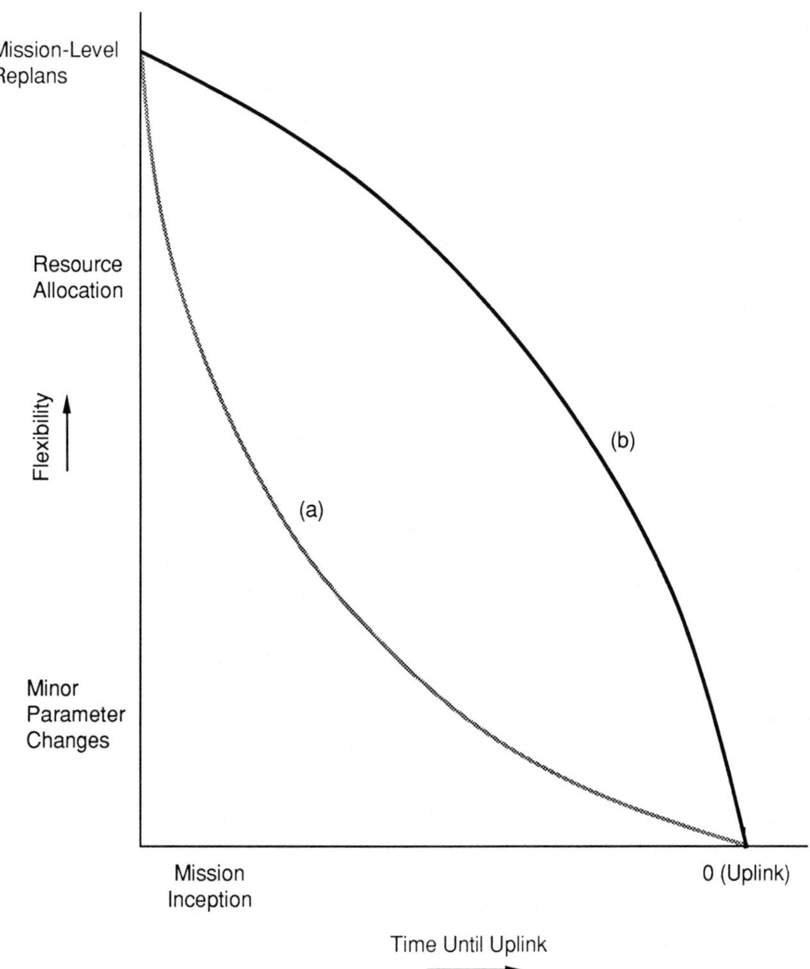

Figure 4.4 The relationship between flexibility and time-to-uplink. Curve (a) represents a lower-cost mission operation than (b).

4.2 Sequence design

Sequence design is the task where the activity plan is expanded into command sequences, which must then be constraint checked and refined for efficiency. If mission planning defines the 'what' of the uplink process, sequence design determines 'how'. The sequencing process assumes that the activity plan satisfies all users and therefore attempts to implement those requests using instructions that platform and payload understand. These instructions can take the form of grouped commands called *sequences*. These are usually marked with the time of execution and stored in on-board computer memory, but they can also be one or more single commands which will execute immediately. Sequence designers must be intimately familiar with the design, operation and capabilities of both platform and payload, for they will determine the specific commands each subsystem needs to satisfy the specified activities. Designers use software that models different parts of the system with varying levels of precision to verify that the developed sequences will perform the intended activities.

If the mission planning task has been done well, most entries in the activity plan can be successfully sequenced. However, even in a well-planned activity there will be unresolvable problems in the design of sequences, and in such cases portions of the activity plan must be redone. Multiple iterations are sometimes required between sequence designers and mission planners. The more complex the platform, the more the possibility exists for deep-seated conflict that remains hidden at the mission planning level, and thus the more likely such iteration will be.

In a perfect world, the only changes that should be accepted in the sequence design process would be 'make play': changes which, if not implemented, would fail to execute the activity plan. In practice this is seldom the case, and user inputs do occur that are too late for inclusion in the activity plan and must be input directly to the sequence-design task. These can be the result of adaptivity, anomaly response, or field coordination (see Chapter 6), but as often they are merely late inputs considered by project management to be worthy of addition in spite of some increase in risk. It is common, too, that some parts of command sequences must be generated late in the uplink process. For example, the Magellan project must use recent (i.e., less than 150 hr-old) orbit determination solutions in order to generate parameters needed by the synthetic aperture radar (SAR). Only these late solutions can predict platform location to the accuracy required by the SAR to set its pulse timing correctly. Once generated, these parameters are inserted into the sequence design task, which is already in progress. A similar problem was faced by Voyager when ephemerides of the satellites of Uranus were not known well enough to aim narrow-angle cameras at them until late in the uplink design process. The SIR project similarly requires late orbit updates to calculate instrument parameters; it also requires last-minute data on sea states and other target properties to properly set its receiver gains.

A useful solution to all of these problems is to design the command system to treat these variables as parameters in commands that can otherwise be constructed early. In the SIR and Voyager cases the parameters are inserted into the command loads late in the uplink process. For its Uranus encounter, Voyager also designed movable blocks — consecutive observations assembled into a separate table whose start time could be varied by real-time command to the spacecraft until 36 hr prior to execution (Morris, 1986). Magellan uses the late data to build tables which are inserted into the uplink stream prior to sequence translation. On the spacecraft, these tables are accessed by on-board computer software to build the actual SAR commands as it sends them to the payload, and to adjust the timing of SAR command execution with respect to orbit periapsis. The Magellan solution has the added advantage that no incorrect table entry can create an unsafe situation on the platform, whereas the others create the possibility (albeit unlikely) of sending an incorrect command.

The complexity of the sequence-design task varies strongly with the complexity of the mission, and especially with the complexity of the platform. An important factor in the design of the uplink process is the presence of on-board data storage. For a variety of reasons it is often advantageous to have the ability to store data on board the platform. When telemetry links are not continuously available, or when the instantaneous rate at which data must be acquired exceeds the capacity of the telemetry channel, on-board storage can divorce data acquisition strategies from telemetry relay requirements and thus offer increased flexibility in planning. However, that capability creates the need for careful design of the strategies which will take advantage of it. For example, if the data storage device is a tape recorder, the ground crew must keep track of the position of the tape with respect to the record/playback head and know the data content on various tracks at any time. Without careful strategy design, the advantages will be quickly outweighed by the complexities created in both uplink sequence design and downlink data processing. Similarly, many space-qualified tape recorders do not allow fast forwarding or rewinding capabilities. Therefore, sufficient time must be allowed to not only replay all recorded data, but also to reposition the tape heads for the next recording. For this reason, tape recorder strategies which reverse the direction of recording on alternate tracks are employed, as the following example will demonstrate.

A useful tool to aid in tape recorder management is a tape map (Figure 4.5), which has been used by Magellan, Viking and other projects to track the position of data on various tracks of on-board tape. Many tape recorders (and most other types of sequential-access storage media) can record and playback in both directions, and it is often advantageous to take advantage of that fact in designing tape management. In Figure 4.5 a set of Magellan star calibrations acquired from day-of-year (DOY) 155 to 169 has been recorded on to track 1 of a tape recorder (here referred to as the data management subsystem(DMS)) beginning 8470.92 in from beginning-of-tape (BOT). The recorder was switched to track 2 and continued to record star calibrations on DOY 170 to 175, ending at 16738.14 in from BOT. Note that the Magellan tape management includes computation of cumulative feet of tape passed across the tape head, stop–start cycles, and cycles of a spring within the tape mechanism. These items represent consumable resources that could lead ultimately to failure of the tape recorder. Final tape position X_F is carried forth to the next tape map. During playback of the star calibration data the tape could be forwarded (on track 2) back to 8470.92 in, switched to track 1 and played, but less head wear would result if the data were re-played

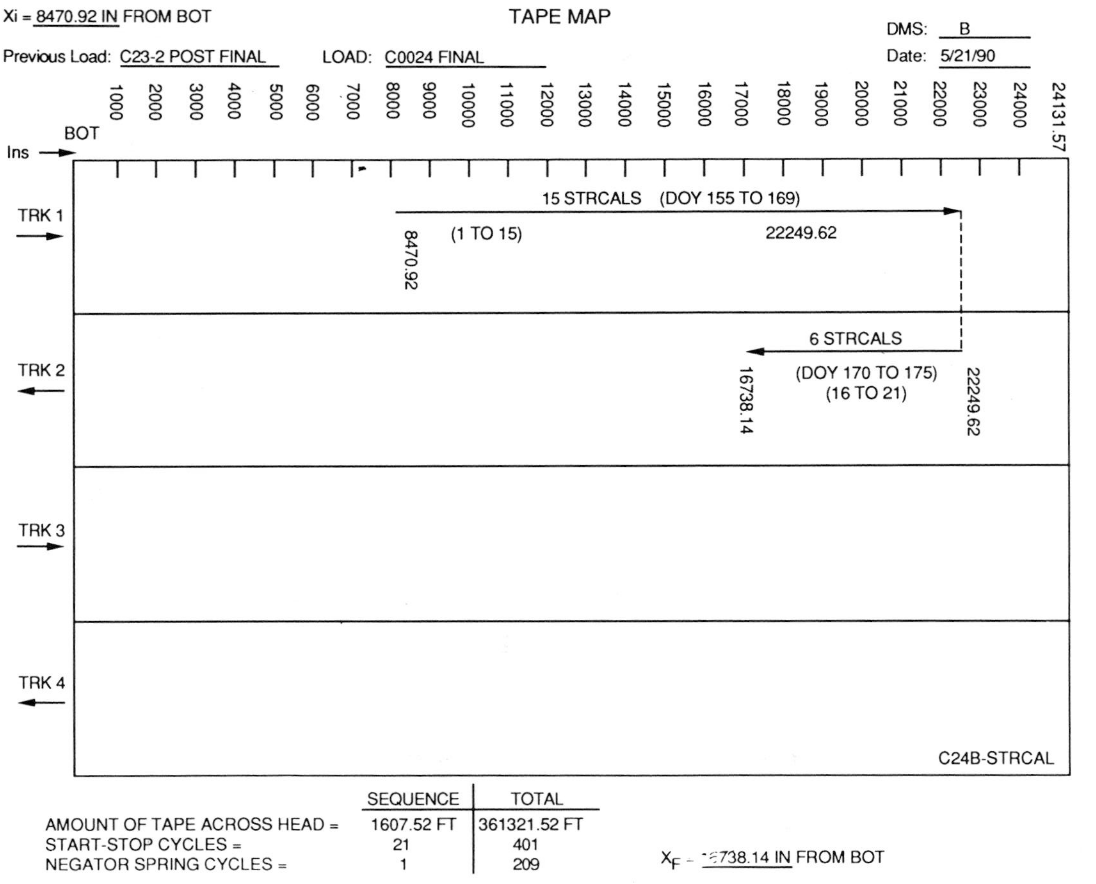

Figure 4.5 A Magellan tape map.

Table 4.2 Parameters of importance in tape management

data content
playback direction
playback speed
record direction
record speed
recorder track number
tape position
time of data playback
time of data recording

backwards from track 2 and continued backwards to the start of the recording. Note, however, that this practice requires the downlink system to be able to decommutate data backwards. Table 4.2 lists important parameters to remember as a part of tape management.

In addition to taking the activity plan as input, the sequence design task also requires accurate documentation of platform and payload command structures, definitions and legal ranges for command parameters required to perform specific actions. From this input, five steps are performed. In the first step, a data management strategy is developed which addresses each line of the activity plan. Downlink opportunities are interleaved with data acquisitions, and each acquisition is planned for either real-time downlink or recording onboard depending on available coverage by ground stations. Playbacks of recorded data and other data routing activities are added to the activity plan and to the sequence as needed. Regardless of the actual data storage medium, the basic purpose of this step is the same, although the detailed constraints will vary. In the SIR-B example of Figure 4.2, the medium is removable magnetic tapes, and some of the basic constraints are as follows:

(1) A datatake may not be split between tapes.
(2) Maximum tape content is fixed and may not be exceeded.
(3) Once filled, a tape may either be: (a) removed and saved for playback after landing, (b) played back to ground during an available telemetry pass and then reused, or (c) removed and remounted later for playback. Priority during telemetry passes is for 'real-time' datatakes.
(4) If taped data are played to ground, tape 'runup' should be accomplished prior to the pass wherever possible.

Such constraints are, of course, highly individualized by mission. More complex platforms also require the assignment of on-board data-routing networks, ground terminals, or ground tape systems in this task.

In the second step, each activity is expanded into the detailed events necessary to accomplish it. A given activity may require many events to implement. In performing this step, all of the instructions to manage the

Table 4.3 Detailed activities scheduled during sequence design

Payload

- sensor on/off sequences
- sensor parameters (shutter speeds, gains, pointing, filters, etc.)

Platform

- sensor pointing systems (star sightings, reaction wheel unloading, etc.)
- on-board data routing
- data storage device (e.g., tape recorder) star and stop
- transmission systems

Data take	Orb	Data Sink	Start time (d/hhmm)	Stop time (d/hhmm)	Attitude	Look Angle	Calibrator	cal level	gain
92.2	92	op	5/1553	5/1605	BLU	50.0	off	2	4
92.4	92	op	5/1615	5/1622	BLU	40.0	off	1	3
92.6	92	td	5/1625	5/1629	BLU	40.0	on	1	3
D92.6tape		plybk	5/1632	5/1640					
92.8	92	td	5/1705	5/1721	BLU	31.0	on	1	2
93.2	93	td	5/1732	5/1738	BLU	18.0	off	0	1
93.3	93	tp	5/1748	5/1800	BLU	50.0	off	2	4
93.4	93	op	5/1805	5/1810	BLU	50.0	off	2	4
93.5	93	op	5/1817	5/1821	BLU	20.0	on	0	2
D93.5tape		plybk	5/1831	5/1841					
93.6	93	td	5/1844	5/1853	BLU	53.0	off	2	4
94.2	94	tp	5/1904	5/1910	BLU	25.0	on	1	2
94.3	94	op	5/1920	5/1929	BLU	50.0	off	2	4
94.4	94	tp	5/1932	5/1937	BLU	40.0	off	2	3
94.6	94	tp	5/1941	5/1946	BLU	57.0	on	2	4
95.2	95	tp	5/2036	5/2040	BLU	38.0	on	1	3
95.4	95	op	5/2047	5/2102	BLU	53.0	on	2	4
crew wake			5/2100						
95.6	95	op	5/2111	5/2115	BLU	53.0	on	2	4
tape offload			5/2130						
maneuver blu-grn			5/2134	5/2150					
96.2	96	tp	5/2152	5/2158	GRN	28.0	off	1	2

Figure 4.6 A SIR-B expanded activity plan.

ancillary systems are added into the activity plan at their proper time. Examples of the level of sequence considered at this stage are shown in Table 4.3.

Figure 4.6 is an expansion of the SIR-B activity plan excerpt shown in Figure 4.2. Note that many columns, such as datatake number and orbit, have been copied directly. Others, such as start time, have been modified as required. To the 'datatake' column have been added other activities, such as tape playbacks or offloads (i.e., removing the tape from the recorder and storing it for later use or for retrieval after landing). Additional columns have been added for instrument parameters, which separate software has calculated based on knowledge of the target and the experiment. 'Cal level' refers to the setting of an internal calibration signal which is injected into the downlink data to be used as a reference when images are produced, and 'gain' is a parameter controlling the amplitude of the received signal before it is digitized on board.

The data destination, or *sink*, for each datatake has been assigned as either 'tp' (indicating that the data are to be recorded using the on-board magnetic tape recorder), 'op' (indicating use of the on-board film recorder) or 'td' (indicating that the data are to be routed directly to the tracking and data relay satellite). These choices are made manually, based on various issues dealing with data quality and availability of the relay satellites. Software which calculates tape usage based on start and stop times has inserted two tape playbacks. Since the Shuttle is a manned platform, full tapes may be removed from the tape drive when the crew is awake. This option has been chosen once in this segment, as indicated by the 'tape offload' at mission time 5/21:30. Note also that mission planning software has recognized the need to change Shuttle attitude at 5/21:34, and has inserted a 6-min manoeuvre accordingly. However, the software recognized that a mission rule required a minimum 2 min interval between any two activities and thus identified a conflicting overlap. Manual intervention was required to decide whether to move the tape offload or to shorten the data acquisition. The mission planner chose to shorten the acquisition and adjusted the start time of the datatake to 5/21:52, as can be seen by comparing Figures 4.2 and 4.6. For this mission, similar rules prevented the requiring of crew activity within 30 min of their wakeup, and required a 4-minute minimum time to accomplish a tape offload before the manoeuvre activity if the same crew members were involved in both.

The third step, expansion of each line in the activity plan into the detailed sequence of commands necessary to accomplish it, is the heart of sequence design. This expansion includes instrument start-up, datataking and turn-off sequences as well as the commands to the platform which are required to support the activity. This expansion into commands is made easier and more risk free if groups of commands that perform specific mission functions are handled as blocks. These command blocks can then be built, tested and validated ahead of time and can be used intact in a sequence without the user having to worry about the relative order and time spacing of the commands within the block. Examples of command blocks are those that perform tape recorder playbacks and vehicle turns, cycle through pre-datatake calibrations, and those that perform spacecraft housekeeping functions such as reaction wheel desaturations.

Command blocks can be fixed, so that the specific commands and command parameters that are issued by the flight processor are identical every time, or there can be block options so that the specific commands will differ slightly to take advantage of options available in the flight system. This definition and use of repetitive command blocks is similar to subroutines or procedures as used in software, and the options are controlled by parameters similar to passed parameters in some software languages. Options may exist to select the tape recorder track for playback, or to select a filter or frequency for an instrument data sample. Identification of instructions which can be executed again and again in the MOS design greatly simplifies flight software, because less needs to be written, integrated and tested. It also simplifies command verification, since the blocks of commands are verified pre-launch using various combinations of parameter options. Then such blocks can be used safely without verification at each use. Options implemented with block parameters provide block flexibility and contribute to a more responsive mission. Experience has shown that command blocks built with no flexibility (with few rigid options or none at all), and coded into ground software, are difficult and time-consuming to modify to adapt to changing operational conditions, new science requirements or anomalies. Shorter blocks built to perform single-function tasks, incorporating various options, can be stacked together in many different ways to respond to a variety of in-flight needs. However, if this is done, significantly more testing is required before launch to validate all the blocks and block options and to provide the assurance of safety mentioned above.

The SIR project used fixed blocks called *macros* of a rather small instruction set for its entire A and B missions. The Magellan project, operating a spacecraft with a large command set, uses corresponding large command blocks with numerous options. Magellan assumed slightly higher mission risk by verifying blocks prior to launch with only representative sets of parameters. In addition to their use of command blocks, the Voyager project used repetitive sequences of blocks called *cyclics* as a way of maximizing sequencing efficiency.

Simultaneously with the command sequence to be loaded on board, a human-readable *sequence of events*, or *script*, is generated in either digital or hardcopy form for ground controllers and crew members (if they exist) so that scheduled activities can be carefully monitored as a part of the downlink process. Contained in either the same listing or in a separate one are the ground activities necessary to transmit the uplink itself and to receive any downlink which will occur as a result. Large projects may actually generate such schedules at several time scales and levels of detail so that, for example, project management and public relations representatives may observe the passage of events at a high level while flight controllers have listings of second-by-second detail. Projects which share equipment such as computer facilities or telecommunications with other projects will need to integrate the other's requirements into their schedules as well to provide maximum visibility.

An excerpt from a script generated for SIR-B controllers is shown in Figure 4.7. The excerpt corresponds to two lines of the expanded activity plan in Figure 4.6. Note that the time period covered by the left-hand column corresponds to the planned duration of datatake 92.6 in the activity plan. References to the previous datatake, numbered 92.4, have been deleted from this figure for clarity. The script shows all events which occur related to the datatake, whether the event takes place in the operations centre, within

```
TIME       FROM      TO                 ACTIVITY
d/hhmmss

5/162000   SIR-B     all sta   five minute warn and runthrough, DT 92.6

                                    4:00 data, 45Mb/s
                                    data start 5/162500
                                    data stop 5/162900
                                    tape start 5/162400

5/162300                       command DWP = 0
                               execute block STA (gain = 3, call lvl 1)

                                    addr  cmd
                                    1770 24000     Inh enable,Dig pwr off, Radar pwr off
                                    1771 24000     Inh enable,Dig pwr off, Radar pwr off
                                    1772 04017     gain state 3
                                    1773 14000     callvl 1,ESG scan at 0
                                    1774 34000     sequencer off

5/162400                       command antenna move 50 -> 40
5/162421                       PREDICTED AOS TDRSS EAST
5/162400   GSFC      SIR-B     ground tape start
5/162430   SIR-B     GSFC      ground tape 30-sec warn
5/162449                       execute block STB (PRF=1824, cal ON, high rate)

                                    addr  cmd
                                    1260 24000     Inh enable,Dig pwr off, Radar pwr off
                                    1261 24000     Inh enable,Dig pwr off, Radar pwr off
                                    1262 00157     Hi dr,cal, 5-bit,PRF 1824
                                    1263 37720     jump to 1720 (TDRSSTART)
                                    1720 24000     Inh enable,Dig pwr off,Radar pwr off
                                    1721 24000     Inh enable,Dig pwr off,Radar pwr off
                                    1722 20001     xmtr off,film adv off,trigger on
                                    1723 24002     Inh enable,Dig pwr ON,Radar pwr off
                                    1724 24002     Inh enable,Dig pwr ON,Radar pwr off
                                    1725 24002     Inh enable,Dig pwr ON,Radar pwr off
```

```
                                        1726 24002    Inh enable,Dig pwr ON,Radar pwr off
                                        1727 24003    Inh enable,Dig pwr ON,Radar pwr ON
                                        1730 24003    Inh enable,Dig pwr ON,Radar pwr ON
                                        1731 24003    Inh enable,Dig pwr ON,Radar pwr ON
                                        1732 24003    Inh enable,Dig pwr ON,Radar pwr ON
                                        1733 24003    Inh enable,Dig pwr ON,Radar pwr ON
                                        1734 20005    xmtr ON,film adv off,trigger ON
                                        1735 20005    xmtr ON,film adv off,trigger ON
                                        1736 34000    sequencer off

5/162500                      DATA START DT 92.6
5/162634                      PREDICTED LOS TDRSS WEST
5/162700  SIR-B     all sta   five minute warn and runthrough, plybk D92.6

                                        8:00 data, 30Mb/s
                                        data start 5/163200
                                        data stop 5/164000
                                        tape start 5/163100

5/162850  SIR-B     GSFC      ground tape 30-sec warn
5/162858                      execute block TDRSSTOP

                                        addr  cmd
                                        1740 24000    Inh enable,Dig pwr off, Radar pwr off
                                        1741 24000    Inh enable,Dig pwr off, Radar pwr off
5/162900                      DATA STOP DT 92.6
                                        1743 20001    xmtr off,Film adv off,trigger ON
                                        1744 20001    xmtr off,Film adv off,trigger ON
                                        1745 20001    xmtr off,Film adv off,trigger ON
                                        1746 20001    xmtr off,Film adv off,trigger ON
                                        1747 20000    xmtr off,Film adv off,trigger off

5/162920  GSFC      SIR-B     ground tape stop
```

Figure 4.7 A segment from the SIR-B command script corresponding to two lines of the expanded activity plan shown in Figure 4.6.

the payload, or on the TDRSS communications satellite, in order by the time of occurrence. For distant planetary missions these times would be corrected for communications propagation time (*light time*) so that events which occur on the spacecraft appear in the list at the time where ground controllers would observe them. As many events are announced by one controller to another, columns are allocated to show who announces the event, and who should be listening. Each activity is described in detail in the 'activity' column, including the expected effect of the event.

Protocol for SIR-B required that a verbal notice be given by the SIR-B controller ('SIR-B' in the 'FROM' column) to all concerned stations 5 min in advance of the start of data. On day 5, at time 16:20, the controller announces that datatake 92.6 will contain 4 min, 0 seconds of data which will be downlinked at 45 megabits per second. He announces the data start and stop time as shown, and the time to start the ground tape recorders so that they will have the required 1-minute start-up interval. Automated commanding of the instrument by the onboard sequencer begins at 5/16:23. The instrument parameters are commanded with both discrete commands (data window position, or DWP, and antenna slew) and with a macro titled STA, which sets gain and calibrate level. The individual commands in the block are then listed together with the expected state of the instrument after their execution. Note that these blocks of commands contain the lowest level of events. STA, for example, has safety power off commands (24000) in case it is called incorrectly. At 16:24:00 tape recorders at Goddard Space Flight Centre (GSFC) are started, and a verbal warning of data start is given 30 seconds before data begin. The block STB is executed at 16:24:49, properly timed so that data will start at the required time. The STB macro calls another macro, TDRSSTART, to actually power on the instrument, as shown.

Note that during this datatake a predicted handover from one TDRSS to another is shown, with the eastern satellite acquiring the Shuttle (acquisition of signal, or AOS) at 16:24:21, just before the datatake starts, and the western satellite dropping out (loss of signal, or LOS) at 16:26:34 during data collection. Although such station handovers should be transparent to the operation of the instrument, it would be vital to know when this event is predicted should a problem in scheduling TDRSS East occur.

The expanded activity plan indicates that a playback of an on-board tape is to begin at 16:32, so the 5-min verbal notice is scheduled at 16:27, even though datatake 92.6 is still in progress. At 16:28:50 a warning is issued to the ground tape controllers. At 16:28:58 the block of commands called TDRSSTOP begins execution, terminating the datatake at 16:29 as scheduled. The block allows 4 seconds of trigger pulses beyond the end of data (an instrument calibration requirement) and powers down the instrument. At 16:29:20, 20 seconds after data stop, the ground tape recorders are stopped by GSFC. The events for datatake 92.6 are complete.

After the generation of a command set and a sequence of events is complete, the fourth step of the sequence design task can begin. The expanded command sequence and the associated sequence(s) of events are constraint checked for: (1) inter-experiment conflicts in resource usage; (2) violation of mission rules and mission management policies; (3) violation of limits such as sequencing memory, data storage memory, or downlink capacity; and (4) any violation of command issuance timing. Automated and manual review by payload, platform, user and other MOS personnel is required to ensure that all systems will be able to execute the planned activities, and that the planned activities are still consistent with original requirements. Most major resource conflicts should have been identified and solved at the mission planning stage, but detailed conflicts will often not be identified until all activities have been scheduled and sorted. Errors that have slipped into the command stream also must be ferreted out before the command sequence is approved. This is best done with a line-by-line walkthrough of the command listings in a review with all platform subsystems, systems engineering, instrument engineers, sequence design engineers and the users present. In many cases, this review is supplemented by submitting the developing sequence to various software models, such as one for Magellan that checks to see that the spacecraft turns during radar mapping passes do not violate any gyro turn rate restrictions. Any discovered violation of these constraints forces the sequence to be further reviewed and alternate implementations used. If necessary, planned activities must be modified so that they can be implemented.

Errors discovered are identified, and the desired corrections are considered by a controlled sequence change activity. Since changes which fix detailed problems can create conflicts at higher levels, changes which are made must be checked to ensure that the overall user needs are still being met. For very small projects there may exist a single individual who has broad enough knowledge to fix problems with certainty, but the risk of creating worse problems by making changes is high in projects of any size. The error correction process is usually an iterative one, with the various MOS teams reviewing sequences, finding errors, submitting changes, and re-reviewing sequences until they are error-free.

Conflict resolution, the fifth and final step in the sequence generation task, is performed whenever a conflict is discovered in the expanded plan. The impacts of conflicts are assessed and potential solutions evaluated by both the sequence designer and the affected users or their representatives. As with mission planning negotiation, the resolution of conflicts by simply imposing priorities is usually not the best procedure, and creative solutions which satisfy the intent of all users can frequently be discovered by interaction between sequence designers and users. Changes resulting from the negotiation of conflicts are input to the sequence change control process described above.

Outputs from the sequence design task are a conflict-free sequence ready for translation into computer instructions and whatever sequences of events are necessary for ground controllers to monitor the on-board events and to control ground equipment. Customarily, files are also generated in this task to provide data for annotation of products to be delivered in the downlink process.

It is worth noting at this point that operational flexibility can be a valuable asset, and that attention should be paid to flexibility in the design of interfaces between the steps we have outlined here. As with mission planning, the level of automation of sequence design varies with the complexity of the mission. When there is significant automation, there may be many intermediate files which are passed from one step to the next. These files may contain long lists of events, times, commands, instrument parameters and so forth, and past projects have attempted to save space by implementing them as binary files rather than as text. It is true that accuracy is preserved and space is saved through such implementation, but what is traded away for that advantage is flexibility. Binary files are not easily edited except with custom software, and writing such software is not a simple job. Unforeseen circumstances in which either the automated design software cannot handle at all or cannot handle within the available time can often be accommodated with a text editor or word processor if intermediate files are kept in text form. The savings in operational flexibility can be well worth the loss in efficiency.

4.3 Sequence translation

Sequence translation is the task of converting each line of the sequence into the low-level instructions which the on-board computer will execute to accomplish the activities intended. The process is similar to software compilation, with the minor exception that absolute instructions (instructions referred directly to memory locations) are generated. The translation accesses the *command database*, a spacecraft-unique data file resident in the ground command system which provides the correct binary bit pattern for each valid command and command option. This task then constructs the uplink command frame where the actual command bit patterns are surrounded by appropriate headers and trailers containing information such as command message type, number of elements and an error detection-correction code. This latter code consists of one or more added words in the trailer that make it possible for the flight software to detect uplink bit errors and perhaps correct them if they are few in number.

The sequence translation task is also where flight computer sequencing memory assignments are usually made for those command sequences that will be stored on board for later execution. Most modern spacecraft utilize stored sequences where the commands are labelled (in spacecraft clock counts) for the appropriate future execution time. A flight software function continually checks the time-tag on the next sequential command against the current on-board clock time and issues the command to the designated subsystem or instrument when the time is reached. Since the memory available for stored sequences is always limited by the flight software design and installation, and since there is seldom a limit to the desires of users, memory space is always at a premium. Many times the stored sequence is too long to fit into one contiguous segment of memory and has to be divided into pieces. The *memory management* function of the sequence translation task will create small sequence control programs which tell the flight software how to put the pieces back together again.

Not all commands are stored sequences, however. In most spacecraft, commands can execute directly from the hardware command decoder or out of a command buffer fed by the decoder. This is the preferred method for single commands that are desired to execute immediately upon receipt, such as those to turn on or off various equipment or for spacecraft emergencies. Some vehicles also have one or more special memory buffers for single delayed action commands that are separate from the stored sequence area. For commands not intended for immediate execution, these special buffers are recommended for use to prevent occupying the command decoder which would further prevent an emergency command from being received.

The sequence translation task is highly specific to the particular flight computer and platform characteristics, is bit manipulation intensive, and is difficult to manually verify correctness, all of which points to the necessity for a highly automated function. For recent planetary missions flown from JPL (Voyager, Magellan, and Galileo) the computer program that performs this function is called SEQTRAN. Each mission has its own version of the software.

Projects which have unsophisticated computers on board, or those for which most platform control is done by personnel outside of the project, may be able to combine sequence design and translation into a single step. However, for most missions this is unadvisable because it could lead to wasted attempts to solve lowest-level conflicts and resource violations before the high-level activity conflicts have been adequately resolved. Leaving translation as a separate step allows higher-level activity conflicts to be resolved before dealing with conflicts in activities such as memory management and low-level data routing.

4.4 Command validation

The purpose of command *validation* (the term includes validation of commands and command sequences) is to ensure that the commands sent to the platform and its payloads will perform the intended activities correctly, and without adverse impacts on any flight elements: in short, with no commanding errors. There are several levels to the overall command and sequence validation process, some of which have been discussed. At this point in our discussion of the uplink process, sequences have been put through automated constraint checking; have been subjected to a detailed walkthrough by involved teams; have had components checked by external software models; have had all changes subject to control and approval; and probably have iterated through this command sequencing cycle more than once. None the less, for the same reasons that spacecraft hardware is tested unit by unit and then again with the units assembled into an integrated package, the command load should be validated by execution on a hardware and software system that models the platform and instrument before it is sent to the actual platform. It is worth noting that a surprising number of commanding errors have been made in both small and large projects. Not only have costly losses of data occurred, but missions have been lost as a result of failure to adequately validate commands before sending them.

Command validation is usually accomplished by a model of the flight system, often consisting of a combination of prototype flight hardware, flight software executing on a computer functionally equivalent to the flight computer, and software simulators. The input to the validation step must, of course, be identical to the command file which is to be uplinked. Simulation can be done at various levels. An ideal simulator uses the command load to predict the resulting datatakes of the targets to be sensed by an instrument, the amount of data to be recorded or downlinked, and the performance of the platform during the instrument operation in areas such as the power load, pointing turn rates, and flight computer executions. If it emulates the spacecraft in enough detail, the simulator can identify low-level problems such as computer routine timeouts or data collisions. It can generate a second-by-second listing of events as they will occur on the spacecraft not only for the period intended to be controlled by the command load but also continuing past that period should the following upload not be received on time. The list can be divided into products listing events of concern to various teams. Each of these products can be sent to the appropriate teams for confirmation of correctness: the target acquisition list to the user, tape usage to mission planners, and vehicle performance data to the respective spacecraft subsystem engineers. Each of these can then be carefully checked against original plans and expected states to verify that the command load is correctly built.

The problem with such a simulator is that it is expensive to build and to run, and that verification of its results is time-consuming. If the simulator uses flight-like equipment, it probably cannot predict spacecraft operation faster than real time: prediction of a day's performance takes a day of simulator operation. Thus the simulator must run all the time to keep up with the spacecraft, and it may require a larger staff to validate the output of such a simulator than to watch the actual spacecraft. Worse yet, simulation may take even more than one workday, since the ideal spacecraft simulator must do many things besides model the spacecraft activities. It must make reasonable predictions of the actual spacecraft's environment, such as positions and brightnesses of the Sun and stars that it needs to use for position and attitude calibration, thermal environment and the like. If it is to model the downlinked data it must also know something about the target to be sensed.

In reality, even the largest of projects has made some concessions to cost in its simulations, although no project of which we are aware has been willing to accept the risk of not performing any simulation of commands before issuing them to flight hardware. There are several compromises that past projects have found to be an acceptable trade-off between cost and risk. First, it is important to consider what must be simulated. A payload or platform whose commands specify a limited number of predictable and safe states only requires a simulator that will accurately predict that state. For example, if an imaging sensor is capable only of modes which dependably perform imaging, calibration, and stand-by modes, and if each of those modes is simple enough that it is not likely to cause trouble, then those modes only need to be verified to have been commanded. If each mode of the equipment is complex in itself, or if there are many such modes, or if multiple modes can exist in combination (such as a computer that can perform several activities at once), then it may be more cost-efficient to perform the simulation as a series of time steps rather than as a series of states. Second, the level of simulation accuracy, important though it may be, can be traded off against either the time required to perform the simulation or the power of the simulating system. If computers faster than the flight equipment are used for the simulation, they can do it faster, but then either the software or the hardware must be different than that used in flight, and exact functional equivalence would be difficult to attain.

Two categories of simulators have been used frequently in past missions. A *state model* is used by projects whose on-board system is complex. Such a device is often composed of flight spare parts or breadboards in combination with software models. A state model is so called because it mimics the state of every part of the platform and payload as a function of time. State models offer the additional advantage that they can be used to predict future states of the spacecraft. This capability is a very useful tool. With a state model, the uplink command load under test should be executed in the model not only through the last planned event in the uplink but through the next communication event with ground stations. An example will show why this is so. Uploads are written into on-board memory, typically into locations prescribed within the load. Let us assume that an error within the sequence translation step would result in uploaded instructions being written into locations used by the flight software for storage of data required to establish a link with the ground. The load is run through a state-model simulator only to verify that all instructions are executed correctly, and they are (albeit from the wrong locations). If, however, the flight software which accesses the overwritten memory is not required to execute until after the up-load has completed its last activity, the next downlink will occur incorrectly or perhaps not at all. If the simulator is allowed to run beyond the end of the upload until the next downlink, such an error would be caught.

As we have mentioned above, there are disadvantages to state models. If they are a detailed model of the on-board system, often using identical hardware, they must generally run at or near real time. Thus a two-week sequence will take two weeks to verify — an amount of time not usually available to a constrained uplink process. In order to save time, simulators are typically built with the capability to perform *clock kicks* — a way of skipping long periods of time where activity is low. The inevitable result is, of course, a compromise to sequence validation completeness.

Simpler or more cost-conscious projects verify command loads with a *command interpreter* which describes the action taken as a result of each command. Command interpreters simply output a description of the action which the platform and/or payload will take when the command is executed, without making any attempt to describe internal actions taken by the on-board computer to produce the result. Payloads and platforms whose on-board computers do not multiplex activities nor share data paths are well served by interpreters. However, if on-board systems are sufficiently complex or autonomous that the possibility of internal data collisions exists, or if inadvertent memory overwrites can affect future scheduled operations, fatal errors can go undetected using command interpreters.

Outputs from the command validation step are lists of activities intended for either automated or manual checking against requirements. It is prudent to do both. Command loads should never be uplinked until the validation step is complete and its output has been thoroughly reviewed. It is also prudent to subject single commands to the command validation simulator, even if they are being sent as a response to an on-board anomaly. In the heat of contingency response, errors are more likely, not less, to be made, making use of a validating simulator even more necessary. One saving grace is that the validation output from single commands is usually not very time consuming to verify.

4.5 Command

In the task called commanding, the upload is transmitted to the platform. The command task is highly dependent on the nature of the platform and its distance from Earth, and is generally operated by non-project personnel. Individual projects usually are limited to a monitoring and coordination function and serve to verify that loads are transmitted successfully. The reader is referred to the references for detailed description of spacecraft communication (e.g., Corliss, 1976; Smith and Hunter, 1981).

Earth-orbiting platforms have used a world-wide set of ground antennas called the NASA Satellite Tracking and Data Network (STDN). In the 1980s projects have developed to the tracking and data relay satellite system (TDRSS), which consists of a constellation of geosynchronous satellites which transfer data up and down through a ground station in White Sands, New Mexico. United States extraterrestrial missions send commands using the Deep Space Network, a system of three sets of large antennas located in California, Spain and Australia (see Figure 1.10). The US Air Force uses its own network of remote tracking stations at eight world-wide locations to control its collection of remote-sensing satellites.

Extreme measures are used within these systems to avoid transmission errors in command loads. Conservative data rates and error-reduction coding are used for such transmissions. None the less, zero bit error rates are never fully achievable, and command loads are particularly susceptible to single-bit errors. Thus for crucial commanding and during periods of questionable commanding (such as solar occultations for planetary missions) uploads are loaded on-board well in advance of the required execution and read back to the ground station for comparison with the intended load while there is still time to correct the new load should it contain errors. If a state model simulator has been used, the on-board computer memory can also be read to the ground and compared with the memory of the simulator as an additional safeguard.

The uplink process ends when the prescribed activity is accomplished. Then begins the downlink process, the subject of the next chapter, by which (among other things) the only concrete proof is obtained that the uplink was actually executed properly.

4.6 Summary

In this chapter we have described the process by which broad and sometimes vague goals of science are made concrete and detailed. These detailed plans are then transformed into the commands that will implement them through a set of steps, each examining the plans at a slightly lower level than its predecessor and searching out inconsistencies. The opportunities afforded by the planned mission to satisfy these plans are identified, and an activity plan is assembled. The plan is examined carefully for self-consistency, for consistency with the available resources, and for consistency with established project policies regarding risks and priorities. Each activity in the plan is then expanded by the addition of necessary supporting activities, and conflicts among the new ensemble are identified. When all activities are listed and are conflict-free, sequences are designed which will accomplish each of them. After more conflict identification and resolution, these sequences are expanded into the commands, controller actions, and other events which will have to occur when the plan is executed, and still more review is done. Finally, the sequence is translated into instructions that can be understood

Table 4.4 Key terms from Chapter 4

activity plan	mission rules
command database	opportunity timeline
command interpreter	repetitive sequence (cyclic)
command simulation	Science Steering Group
command validation	sequence (or command sequence)
constraint checking	sequence design
conflict resolution	sequence of events (SOE)
data-take	sequence translation
downlink process	state model
Experiment Representative	tape map
inverse planner	tape recorder strategy
memory management	uplink process
mission planning	

and executed by on-board computers, those instructions are sent to platform and payload, and the events in the plan actually take place. If the task has been done well, the goals are accomplished as expected.

Table 4.4 contains a list of the key terms used in this section. We next turn to the downlink process, where data are processed, checked, and made into products useful for analysis.

4.7 Exercises

(1) The Viking mission, comprised of two landers and two orbiters, required an especially sophisticated tape management strategy. Consider a single lander with a camera having a data rate of 16,000 bits/second and a meteorology experiment which produces a 2000-bit frame once each minute. The camera is to take as many 30-second images as possible to look for moving objects; the preferred interval is once every 15 min. The lander tape recorder has capacity of 50 megabits, and it can dump its recorder to an orbiter via a link that lasts 20 min every 7 hrs. Devise a tape management strategy for a 24-hr period and produce a tape map.

(2) Explain how the increasing capability of flight computers in both processing speed and memory size has made the commanding process more complex and difficult for the ground to implement. Are there any ways in which this increased capability has helped the design of the mission operations system?

(3) Give some examples of how a remote-sensing mission might use:

(a) single commands that execute immediately upon receipt;
(b) single commands that execute at a later time as specified in the upload;

(c) command sequences;
(d) repetitive command blocks.

(4) For an Earth polar orbiter with a mission objective of mapping the seasonal variations of the extent of the polar ice caps, what might be some of the key elements of the mission plan?

(5) A geosynchronous infrared astronomy satellite has two principle instruments that must share time. The first is an infrared spectrometer that can function at three different wavelengths, *w*1, *w*2 and *w*3, and has two different filters (A and B) that can be rotated in front of the aperture one at a time or it can use no filter at all. Spectrometer measurements at wavelengths *w*1 and *w*2 each require 4 minutes for a sample while *w*3 requires 7 minutes. When either filter is used, the exposure times double, and rotating the filter into the aperture takes an additional minute, as does removing it. The second instrument is an infra-red radiometer that requires 5 min of exposure time to obtain a temperature measurement. The angle between the boresight pointing directions of the two instruments is 45° and the spacecraft turn rate on reaction wheels is 15° per minute. Design an activity plan, similar to that in Figure 4.2, for collecting one each of all possible data samples from four objects in the minimum time possible. The objects form a square on the celestial sphere with sides of 60°. For every 2 hr of data collection (both instruments collect at the same data rate) you must schedule a 20-minute tape recorder telemetry dump to a TDRSS satellite. Assume all turns to and away from TDRSS are included in the 20 minutes.

(6) Select a thirty-minute section of your activity plan from exercise 5 above that includes the first few minutes of a tape recorder playback and expand it into a sample command script. Show all spacecraft activity that must occur to support the indicated data collection and transmission.

(7) After a command sequence file has been translated into binary format, what checks and verifications might be performed to ensure that the translation step has produced the correct file for transmission to the spacecraft? For each check, identify a potential error that the given check will catch.

References

Corliss, W. R., 1976, *A History of the Deep Space Network*, NASA Contractors Report CR-151915, Washington, DC: National Aeronautics and Space Administration.

Harris, H., 1984, *SMDOS: SIR-B Mission Design and Operations Software*, D-1081, internal document Pasadena, CA: Jet Propulsion Laboratory.

Linick, T. D., 1985, Spacecraft commanding for unmanned planetary missions — the uplink process, *Journal of the British Interplanetary Society*, **38**, 450–457.

Miner, E. D., Stembridge, C. H. and Doms, P. E., 1985, Selecting and implementing scientific objectives, *Journal of the British Interplanetary Society*, **38**, 439.

Morris, R. B., 1986, 'Sequencing Voyager II for the Uranus encounter', presentation at the AIAA 24th Aerospace Sciences Meeting, Williamsburg, VA, August.

SIR-B Science Team, 1984, *The SIR-B Science Investigations Plan*, 84-3, Pasadena CA: Jet Propulsion Laboratory.

Smith, J. G. and Hunter, J. A., 1981, 'NASA tracking and data acquisition in the 1990s — high Earth orbit and planetary spacecraft support', presentation at the AIAA 21st Aerospace Sciences Meeting, Arlington, V.A., June.

Wilton, R., 1985, Microcomputers in NASA's SIR-B, *Byte*, July 1985, 193–198.

5

The Downlink Process

The downlink process is defined as the portion of the system which begins with the production of a data stream by either platform or payload and ends with the placement of properly catalogued processed data into an archive. Proper amounts of documentation regarding data production and processing and the associated *metadata* — information about the data which allows them to be conveniently accessed and understood — should also be placed in the archive to allow future users easy access to mission results.

Like the uplink process, downlink has made significant changes in the past 20 years. Early Mariner images, for example, were created by dumping raw telemetry data numbers on computer printouts and colouring different ranges of numbers with different crayons. Plate I shows the result, which, although crude by later mission standards, was at that time hailed as a breakthrough. It was possible at last to actually 'take a picture' from space! The need for a more accurate way of depicting images, demonstrated at least in part by the Mariner experience, has led to development of electro-optical film recorders that now have both high fidelity spatial and intensity reproduction capabilities. Similarly, early Soviet Venera (Venus) landers could not separate engineering data from science data in their downlink process. As a result, images from Venera contained stripes of what appeared to be noise in them because the engineering telemetry which was interspersed with image data in the downlink was not removed before the image was created. As we will see in this chapter, later telemetry processors have several different methods of separating different modes of telemetry into component parts, allowing images to be separated from other science data and from engineering data regardless of the order in which they were acquired.

5.1 Data generation, storage and telemetry construction

The downlink process begins at the sensor, where commands generated by the uplink process are received. The sensor responds by executing a datatake and producing *user data*. User data are also referred to as *science data* in this book because the scientist is the end user in scientific remote-sensing missions. Similarly, on-board subsystems generate *engineering data*, both routinely and through command receipt. For most platforms, the majority of downlinked data is user data — generated by the sensors which form the payload for the end users.

Sensors vary widely in the amounts of data they generate. Engineering temperature sensors may take a single 8-bit measurement very infrequently. Imaging instruments, on the other hand, may generate gigabits in a single datatake. To conserve data volume, some instruments (such as seismometers or weather monitors) are triggered by events they observe. In their quiescent mode they may generate no data, or they may be in a low data rate mode. Many sensors contain data buffers so that their instantaneous data rates can vary as needed while their output rate is either constant or allowed to vary at the convenience of the platform.

Most user data contain not only the remotely-sensed information but also *ancillary data*, data relating to instrument pointing direction, gain state, and similar information necessary for interpretation of the science data. Often the command which instructed the instrument to generate the data is replicated as ancillary data in the downlink to more completely describe the datatake and also to aid in troubleshooting. These types of information are generally contained in a preamble or header which is never separated from the body of the data. More general ancillary data which may apply to several sensors (such as navigation information) are downlinked separately and joined to the user data as a final step in the data production process.

Simultaneously (or nearly so), the platform generates engineering data to keep the ground engineering teams aware of how it is performing. Engineering data are typically composed of data from sensors within the spacecraft, such as temperatures, pressures, voltages, and the like. Many spacecraft also contain some remote sensors for engineering purposes, such as star and/or sun sensors to update the knowledge of the spacecraft's attitude. Most engineering data are not routinely commanded but are read out from sensors regularly. Thus the engineering data stream is more highly predictable than the user data. In nominal situations, engineering data occupy far fewer bits than user data, although they usually originate from more sensors with a wider variety.

Once they have been generated, both user and engineering data are passed to a data handling system on the platform where they are assembled, or *commutated*, into telemetry frames. A data commutation map contained in the on-board software will dictate how the various types of data are to be combined into the telemetry. Note that because downlink bits are a valuable resource, it is not generally possible to keep data in even byte (i.e., 8-bit) words: if a measurement only requires 5 bits of accuracy, only 5 bits are used, whereas in terrestrial computing one might be tempted to 'pad' the data with three additional zeroes for ease of handling. Thus, incorrect processing of downlink data can result not only in the wrong data being assigned to a measurement, but also in 'bit-shifted' data, which can be very difficult to interpret.

Sometimes a platform's radio subsystem will transmit separate telemetry streams for the sensor and engineering downlink (either at divergent frequencies or on different subcarriers of the same frequency), but for most missions the two are combined into a single stream. The Magellan mission actually does both. Commands produced in the sequence design process define whether the data handler is to pass the data directly to the telecommunications subsystem for immediate downlink or to record them in an on-board data storage device for later transmission.

There are several basic reasons for recording data on board. First, communications channels to the ground station are usually not continuous. In the case of the NASA TDRSS, which services many Earth-orbiting satellites, competition for service is often intense, and transmission opportunities must be scheduled at TDRSS convenience. NASA's other major data paths, the STDN and the DSN, are also shared by many users. Most telemetry and command systems have operations systems of their own, designed for multimission use, and individual platforms are their users. But even with a dedicated telemetry and command system, such as some military spacecraft have, orbital or celestial geometry often prevents continuous downlink. Magellan has large blocks of continuous coverage by the DSN; but it is occulted from view at one point during many of its orbits when it passes behind Venus as seen from Earth. The Sun also blocks transmissions during solar occultations. Similarly, the Viking landers had very limited transmission opportunities directly to Earth. One of the primary functions of the Viking orbiters (after they had delivered the landers) was to mitigate this situation by serving as a transmission relay. The higher position of the orbiter gave it much longer periods in view of the Earth. Earth orbiters often are simply out of view of any antenna for periods of time. TDRSS can provide global coverage of most common orbits (if service can be obtained), with the exception of a small area over the Indian Ocean. In other cases, the spacecraft itself is sometimes unable to communicate with Earth because of power, pointing or other limitations. Magellan uses the same high-gain antenna for its SAR and telemetry functions; when it is taking SAR data it cannot use that antenna for transmissions to Earth.

The second function of on-board data storage is to change the data rate. The data rate to Earth is principally limited by platform antenna diameter, although other factors such as available power are also important. Since antenna diameter is in turn limited by the amount of mass or size that can be accommodated by the launching vehicle, data from high data-rate instruments must often be slowed down for relay to the ground. Most tape recorders and other forms of data storage are capable of recording and replaying at different speeds. This feature allows short periods of data acquisition at high rates and, independently, longer periods of playback over lower rate telemetry links.

A third reason for using on-board storage is data consolidation. When a variety of sensors occupy the same platform, they often need to operate at different rates and independently in time. A lander may need to take images when the illumination angles are optimal; it may also need regularly-spaced meteorology measurements, and its seismometer may need to deliver data at irregular intervals. On the other hand, the lander data-link opportunities need to be tied to the positions of Earth and a relay orbiter. Even independent of required data rate changes, the incoming data must be buffered for transmission, or perhaps held for a later down link opportunity.

A difficult trade-off during mission design is whether or not data are to be compressed for downlink. *Compression* is the process by which the number of bits used to transfer information is decreased from that collected. Compression can be either *lossless*, in which case no information is lost in the process, or *lossy*, in which case some information of lesser importance may be truncated. Data bits are precious, and extreme care is taken to be sure that every possible redundancy is removed from data before they are transmitted by lossless compression. Even after all the redundancy is removed, however, information loss is sometimes permitted by further compression in order to save room in the downlink. Much theoretical work has been done in the past two decades to maximize the amount of information carried by a given number of bits (e.g., Shannon and Weaver, 1964).

While engineering data are usually not very compressible, user data often are. This is especially true of imaging data. Images are two-dimensional data sets, and in the process of transforming them into imaging data they are dissected into individual lines. There is often a relationship between picture elements in both directions, and many data compression schemes take advantage of this relationship to remove data from the image and then restore them in ground processing. It is possible to see as high as 10:1 savings in the number of bits required to transfer imaging data without loss in information content. Images of large portions of planetary bodies, such as those produced by Voyager, often contain large black areas representing deep space. Such areas contain little or no information and can be even further compressed. Other data types which contain large amounts of noise are not as amenable to compression. Similarly, SAR data are not easily compressed because of an inherent noise-like characteristic known as speckle. However, in most data, including SAR, some savings are able to be realized as a result of data compression (Kwok and Johnson, 1989).

Despite the obvious benefits to be gained in data compression, there are negative sides to this trade-off. First, data compression on board is expensive in terms of time, power, computer size, software complexity, and relative risk. Second, the decompression required in ground data processing is also expensive. Third, lossy compression can inadvertently delete scientifically useful information. Fourth, losses during downlink telecommunications are more significant since fewer bits are carrying the same information.

The last item raises another issue, applicable to both compressed and uncompressed systems, where the data must be transmitted over large distances through various mechanical and electronic equipment. Transmission of data over radio systems is imperfect, and some percentage of the bits sent out from a spacecraft are corrupted by noise. An important requirement on the link between the platform telecommunications system and the receiving ground station is the allowable *bit error rate*, the statistical maximum percentage of bits that may be incorrectly received. It is not unusual for this requirement to be as low as 1 in 10^6, which means that an average of only one of every million bits may be corrupted. In order to achieve such rates, data must be *encoded*. Encoded data are protected against bit errors through an added level of redundancy. Encoding schemes may add additional bits based on the *parity* of segments of the data bit pattern, which is the oddness or evenness of the sum of the ones in the data segment.

5.2 Data receipt and synchronization

The techniques involved in the receipt of the electromagnetic signal from a spacecraft and conversion of that signal into a data stream involve some of the most sophisticated signal processing known. The power output from spacecraft is very low compared to ground transmitting stations, and for vehicles at planetary distances, the amount of the power that falls on the receiving antenna is measured in billionths of a watt or less. A detailed description of that process is beyond the scope of this book. Broadly, signals are acquired by large dish antennas which are aimed using predictions of platform location provided by individual projects. The large dish structures focus the incoming signals onto cooled detectors where they are converted into electrical currents, which are amplified and fed to a baseband amplifier, from which it is digitized.

Downlink data flow is summarized in Figure 5.1. Data from the receiving antenna are usually irreplaceable (although they can occasionally be replayed from platform data storage devices), so they are recorded as soon as possible. In some cases the data are written to a permanent archive even before digitization, and the analog baseband recording is saved until the validity of the digitized version is proven. But in all cases the digital data are preserved in an *original data record* (ODR) at the antenna site.

As the receiving antennas collect the spacecraft data, they also provide another valuable function used in determination of the platform trajectory. There are several ways in which this is done. Obviously, the direction from which the signal comes roughly determines the right ascension and declination of the spacecraft, but this information is very low resolution and is rarely useful. More important is the exact frequency of the signal, because the frequency is shifted from that transmitted due to the Doppler effect, and the magnitude of that shift is determined by the radial component of the spacecraft's velocity. Taken together, this information is called *tracking* data. It is transferred to the mission navigators, who use it to determine the

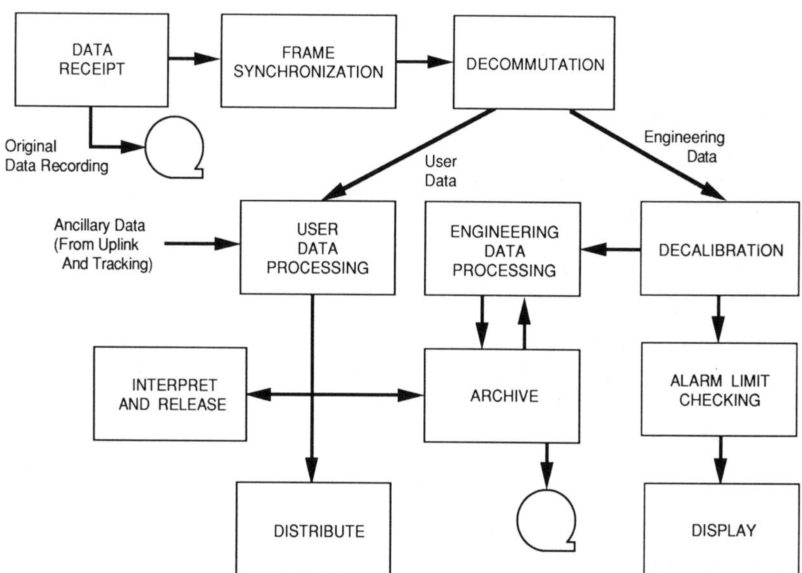

Figure 5.1 Downlink data flow.

spacecraft trajectory. More precise tracking can be achieved through the simultaneous use of two or more ground antennas and a technique known as *very long baseline interferometry* (VLBI).

Downlink telemetry is self-clocking. The single analogue signal must be divided into individual bits by some characteristic of the signal itself, rather than by some external reference to time. Similarly, the division of the data into fixed series of bits that can be recognized (called data *frames*) must be accomplished by looking at the data. An important function following data receipt, then, is data frame synchronization, the determination of the location of the beginning and ending of each frame of data. First, the analog signal is broken down into bits (digitized), and then a search is begun for a fixed, recognizable pattern of bits called the *synchronization word* or 'sync word' that indicates the beginning of a frame. These patterns are chosen carefully so that they are least susceptible to bit errors.

For many past missions, frame synchronization was not accomplished until the telemetry had been transferred to the final data-processing site. The ground stations would segment telemetry into an arbitrary but fixed number of bits (usually about 2000) and insert them into a ground transport block with applicable header information appended, and transmit them through either a communication line or by physical shipment of tapes to the processing site. The added header information contained such auxiliary information as Earth receipt time, processing equipment string identifiers, received signal to noise ratio, and antenna gain during the reception. Lately, however, there has been progress toward moving the frame synchronization function out to the antenna site so that when the data are received at the data-processing site, the beginning and ending of each spacecraft data frame is delineated with End-of-Record marks.

One might think that in a telemetry stream, once the beginning of the first data frame is found, the rest are easily found by just counting forward the number of bits per frame and inserting another end-of-record mark. In practice, however, the situation is more complex. Some spacecraft intersperse data frames of varying length in the downlink. Others permit variable-length frames. Even if all frames were generated the same length, dropouts during the downlink would cause multiple reinitializations of the frame synchronizing algorithm. To avoid loss of data, every frame must contain an identifier in the unsynchronized data. A software process known as *flywheeling* has been developed to take advantage of the prior knowledge of the frame lengths to avoid having to search each bit position and to increase the speed for accomplishing synchronization. In this process, the search is started just prior to the estimated position of the next frame identifier and is continued until the expected pattern is located.

5.3 Decommutation and decalibration

After the data frames have been identified and marked in the telemetry stream, the next step in processing is to extract from the frame the various types of information contained in it and sort them into data bins, through a process called *decommutation.* After being sorted, the extracted data words are converted from data number format to more familiar engineering units (EU, often eu) (such as voltage or degrees) through a second process called *decalibration.*

Decommutation is the ground process that reverses the commutation process on board the spacecraft by taking apart the data frame into its constituent components. There are several ways of designing and implementing the commutation-decommutation process. Sometimes a single data frame is packed with data words of varying types, either science or engineering or both, and this frame simply repeats so that a given measurement can be extracted from the same location in each frame. In other designs, there are several different types of data frames each composed of one type of engineering or user data, and the telemetry cycles through all available frame types in a regular rotating pattern. Here, one complete cycle of all data frame formats is generally called a major frame, while each individual format is called a minor frame.

The best method of describing decommutation is by example. Figure 5.2 illustrates a simplified engineering telemetry data frame for a sounding rocket whose 30-minute repeatable mission is to ascend to the upper atmosphere and sample its content during flight. Its sole instrument is a spectrometer,

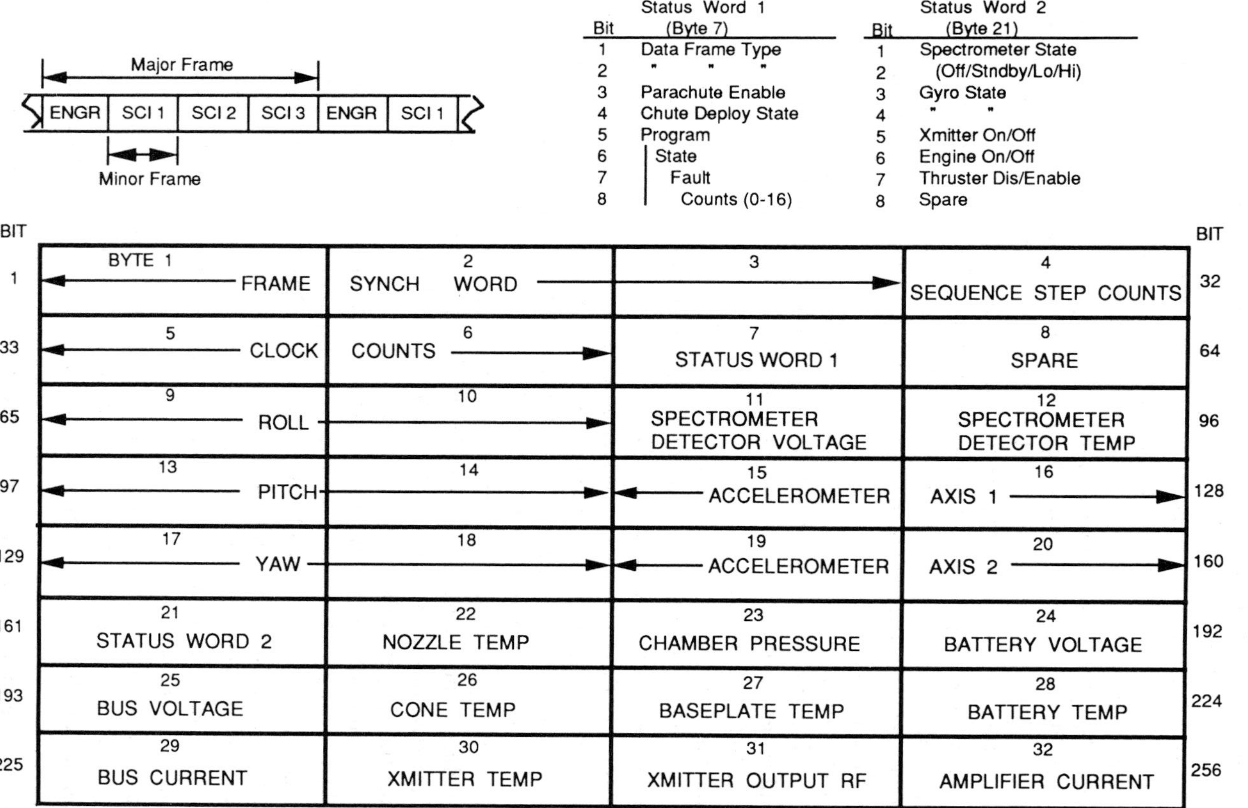

Figure 5.2 An example of an engineering data frame.

which acquires data in the 0.4 to 2.5 μm range in 72 channels. In this example there is one 256-bit engineering data frame and three different user data frames, also 256 bits long. The telemetry commutates through all four data frame types each second during flight, using a downlink data rate of 1024 bits/s. Various engineering data measurements are loaded into the engineering frame and the collected spectrometer data are loaded into the three user data frames.

Each rectangle in the engineering frame of Figure 5.2 represents one 8-bit byte of telemetry. The first three bytes constitute the 24-bit frame synchronization word, which is identical to the frame synch word in the science frames. Bytes 5 and 6 are the clock counts from the on-board oscillator, which puts out 32 counts/s, or one count every 31.25 ms. With a 16-bit integer clock word, the maximum number of counts is 65,536, which will result in clock roll-over in 34.13 min, compatible with the maximum duration of the flight.

Fourteen of the measurements are 8-bit integer words, giving a range of data numbers from zero to 255, for a quantization accuracy of 0.0039 times the full range of the measurement. For example, if the bus voltage measurement in byte 25 can have a range of 0–10 volts, each change of data number by one will result in a value change of 0.039 volts. Five of the measurements, specifically the rocket attitude (roll, pitch and yaw) and the accelerations, are 16-bit floating point words containing a mantissa and an exponent. Two of the bytes, numbers 7 and 21, are status words composed of several one-to-four bit integer values, the content of which is indicated in the upper right portion of the figure. Many of these are single bit words giving the on/off status of certain on board components. The first two bits of status word 1 indicate the type of frame, 00 for engineering, 01 for the first science frame, 10 for the second science frame and 11 for the third science frame. Similarly, the first two bits of status word 2 indicate the four states of the primary instrument, off, standby, low and high.

An example of the accompanying science data frame is shown in Figure 5.3. The frame synch word is repeated, and the time at which the frame was taken is shown in bytes 5 and 6 as before. Byte 4 is a status word similar to those in the engineering frame. The data frame type identifier is the first two bits of byte 7 (as it was for the engineering frame) packed together with parameters relating to the spectrometer. The remainder of the frame is taken up with the spectrometer channel identifiers and data values. Twenty four complete channels are represented in each frame. The first 7 bytes in the other two minor frames have the same definition as in the first, and the remainder of these other frames contain the remaining 48 channels of user data.

Decommutation extracts each measurement from its location in the frame according to a decommutation map, identifies the measurement with the time extracted from bytes 5 and 6, and puts it in a file consisting of time-tagged values from only that measurement.

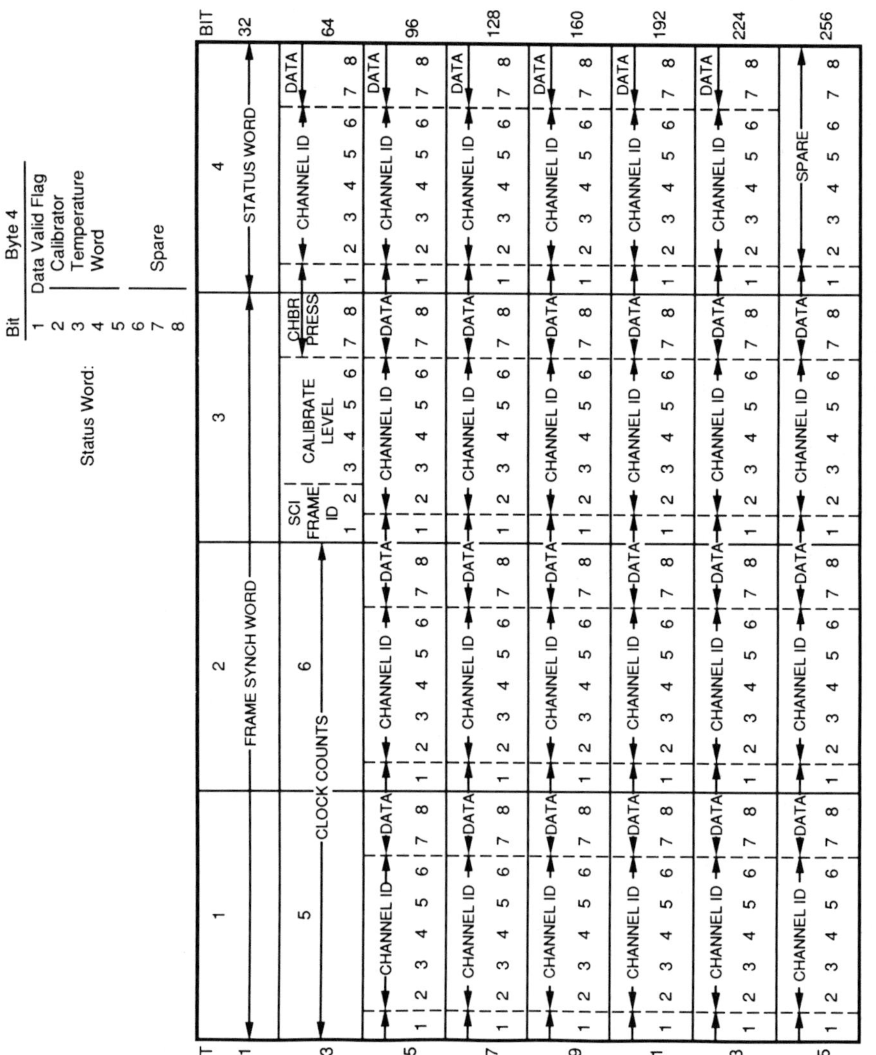

Figure 5.3 An example of a science data frame.

PROG X DECOM.MAP VERSION 4.2 CREATION DATE = 07:14:1990 TIME = 21:23:46
MAP SEQUENCE NUMBER = 0 FILE NAME = XDECOM.MAP42D

FRAME ID	MEASUREMENT	PARENT	OFFSET	SIZE	TYPE	CAL CURVE	UNITS
ENGR	FRAME SYNCH WORD	FRAME	0	24	INTEGER	NONE	NONE
ENGR	SEQ STEP COUNTS	FRAME	24	8	INTEGER	NONE	COUNTS
ENGR	CLOCK COUNTS	FRAME	32	16	INTEGER	NONE	COUNTS
ENGR	STATUS WORD 1	FRAME	48	8	STATUS	NONE	NONE
ENGR	DATA FRAME TYPE	WORD1	0	2	INTEGER	T8.1	NONE
ENGR	PARACHUTE ENABLE	WORD1	2	1	INTEGER	NONE	NONE
ENGR	CHUTE DEPLOY STATE	WORD1	3	1	INTEGER	NONE	NONE
ENGR	PROG STATE FAULT CNTS	WORD1	4	4	INTEGER	T8.2	NONE
ENGR	ROLL ANGLE	FRAME	64	16	FLOAT PT	F9	RADIANS
ENGR	SPECTR DETECTOR VOLT	FRAME	80	8	INTEGER	N11	VOLTS
ENGR	SPECTR DETECTOR TEMP	FRAME	88	8	INTEGER	N12	DEGREES C
ENGR	PITCH ANGLE	FRAME	96	16	FLOAT PT	F13	RADIANS
ENGR	ACCELEROMETER AXIS 1	FRAME	112	16	FLOAT PT	F15	RAD/SEC
ENGR	YAW ANGLE	FRAME	128	16	FLOAT PT	F17	RADIANS
ENGR	ACCELEROMETER AXIS 2	FRAME	144	16	STATUS	F19	RAD/SEC
ENGR	STATUS WORD 2	FRAME	160	8	INTEGER	NONE	NONE
ENGR	SPECTR STATE	WORD2	0	2	INTEGER	T21	NONE
ENGR	GYRO STATE	WORD2	2	2	INTEGER	NONE	NONE
ENGR	XMITTER STATUS	WORD2	4	1	INTEGER	NONE	NONE
ENGR	ENGINE STATUS	WORD2	5	1	INTEGER	NONE	NONE
ENGR	THRUSTER DIS/ENABLE	WORD2	6	1	INTEGER	NONE	NONE
ENGR	NOZZLE TEMP	FRAME	168	8	INTEGER	N22	DEGREES C
ENGR	CHAMBER PRESSURE	FRAME	176	8	INTEGER	N23	PSI
ENGR	BATTERY VOLTAGE	FRAME	184	8	INTEGER	N24	VOLTS
ENGR	BUS VOLTAGE	FRAME	192	8	INTEGER	N25	VOLTS
ENGR	CONE TEMP	FRAME	200	8	INTEGER	N26	DEGREES C
ENGR	BASEPLATE TEMP	FRAME	208	8	INTEGER	N27	DEGREES C
ENGR	BATTERY TEMP	FRAME	216	8	INTEGER	N28	DEGREES C
ENGR	BUS CURRENT	FRAME	224	8	INTEGER	N29	MILLIAMPS
ENGR	XMITTER TEMP	FRAME	232	8	INTEGER	N30	DEGREES C
ENGR	XMITTER O/P RF	FRAME	240	8	INTEGER	N31	DB
ENGR	AMPLIFIER CURRENT	FRAME	248	8	INTEGER	N32	MILLIAMPS

Figure 5.4 The decommutation map for the data frame of Figure 5.2.

A decommutation map for the engineering data frame of Figure 5.2 is shown in Figure 5.4; a similar map would be used for the science frame. For each parameter in the commutated frame one entry is made in the decommutation map which describes how the frame is to be split. First is the frame synch word, offset by 0 bits from the start of the frame. It is 24 bits long and is to be read as a raw integer. Decalibration of this parameter is not required and there are no associated engineering units (i.e., it is a pure number). The 'sequence step counts' parameter is offset by 24 bits from the start of the frame, is 8 bits long, and so on. The 'parent' column identifies the origin of the offset — for example, 'parachute enable' is offset by 2 bits from the start of status word 1. Note that those parameters which are not pure numbers, such as 'spectr detector volt', have a calibration curve file identified (*N*11 in this case) where the curve to be used to convert data counts into engineering units is specified, usually as the coefficients in a polynomial expansion. The decalibration process converts the data number to engineering units. For example, a data number of 129 (bit pattern 10000001) for the bus voltage in byte 25, with the ranges mentioned in a previous paragraph, would linearly translate to a value of 5.039 volts.

The decommutation scheme for actual missions is usually much more complicated than that described above. This is perhaps best illustrated by the Viking lander data stream. There were six different engineering data frames for different mission phases and different activities, containing not only the vehicle's subsystem engineering telemetry, but also data for four of the science experiments. Then, each of six other experiments had their own dedicated data frame. The on-board commutation process interleaved frames for engineering, biology, seismometry, x-ray fluorescence, gas chromatograph-mass spectrometer, imaging, and meteorology into one data stream. All data frames were of differing lengths, so that the frame synch algorithm was constantly searching for frame starts. Moreover, the seismology frame was of variable length, so that if a marsquake were to occur, the frame would expand in size for the duration of the quake.

Difficulties have grown out of past variability in the formats and schemes for building and decommutating telemetry frames, both between missions and within a mission. As a result, there has been an effort toward *packetized telemetry* over the last several years. This concept promotes a standard telemetry frame with consistency in such normal variables as frame size, synch word size and location, clock word, data rates, and subcommutation techniques. Data from each mission would be required to be stored in the frame in pre-defined packets which would have a standard frame size even for different spacecraft. If the space provided in the packet is not sufficient for all the various measurements required by the mission, one or more packets can be designated as *variable packets*, which *subcommutate* several different measurements in the same location over a span of multiple data frames. For example, ten different measurements can occupy the same 8-bit location in a frame if each measurement is only inserted by the flight system into every tenth frame on a rotating basis. The sampled frequency of these values are, of course, only 10 per cent of that of the rest of the telemetry.

5.4 Display

Once the telemetry has been decalibrated into engineering units, it is ready for presentation to the controller or analyst. Although the telemetry may be processed in various ways over many subsequent days, it is usually first presented to human view on a display device in near real time to allow for monitoring of the health of the vehicle and its subsystems (or for quick-look data interpretation). In earlier missions, the display devices were digital television monitors with displays driven by a central computer system. The formats of these displays were generally hard-coded into software and not controllable by the user. The only user option was to select from several different fixed-format displays through push buttons. With the advent of smaller and more powerful self-contained computers, the control of telemetry displays has moved out to the end user, and individual workstations can be reprogrammed in real time to display graphs or other custom displays from the telemetry stream.

A typical modern spacecraft control room contains many small computers connected to a local area network through which live telemetry is supplied. This telemetry may be raw data frames, leaving decommutation and decalibration to be performed in the user's workstation, or it may be provided already channelized by a central computing system. Features commonly available to the user include multiple, independently-controlled windows, list (scrolling) displays, matrix (fixed position) displays, channel-versus-time plots, channel-versus-channel plots, colour displays, mouse or light-pen accessories, and touch screen capability. The ability to call up and display predicted channel plots and overplot actual telemetry as it arrives is very useful. The capability to manipulate the data and build interface files to offline or commercial software such as spreadsheets or database programs is also valuable.

5.5 Alarming

One of the primary uses of real-time (or near real-time) displays is to monitor the health of the spacecraft (or instrument). Since recent advances in on-board computer memory, storage, and processing capability have resulted in increasingly complex telemetry structures, the ability of the display system to quickly call the attention of the user to potential problems becomes of paramount importance. This ability is called telemetry *alarming*, and the process of determining which specific values in a telemetry stream require attracting the attention of a controller is termed *alarm limit checking*. At some control centres the resulting alarms are called *caution and warning* indications. This processing step is normally accomplished immediately after decalibration and before the data are routed to the display devices.

Both high- and low-alarm limits are typically developed for those measurement channels the user identifies as desirable to monitor. The source of the limit values is usually either vendor qualification test data for the flight hardware involved, or environmental testing of the spacecraft prior to launch. Sometimes limits may be derived based on prior flights of similar components. Software-generated measurements may have limits that are simply computed based on potential fault scenarios. Alarms may be triggered by passing either above or below a limit value, and can also be designed to be exclusive of certain values. Alarms can be based on decimal, octal, hexadecimal or binary values, or on alphanumeric matches or mismatches.

The methods of user notification can also vary. Although the most common technique is a reverse video display of the offending value, other ways such as blinking display, dump to an alarm printer or audible alarm have been used. The triggering of an alarm can be designed with *hysteresis*, which will trigger only if the alarm limit is exceeded by a consecutive number of samples, as controlled by the user. This avoids an erroneous alarm due to a bit error in a single frame of data. Alarms can be designed to remain in alarm once the limit is exceeded, or to cycle in and out of alarm as the measurement value does. The former prevents an alarm due to a measurement which exceeded its limit for a short period of time from going unnoticed, since personnel action is required to clear the alarm from the display.

5.6 Engineering data processing

There are two categories of *engineering data* processing applicable to nearly all remote-sensing satellite missions. The first, telemetry analysis, is data processing which permits the engineer to analyse received subsystem telemetry to understand the historical trends of the spacecraft subsystem. The second, performance prediction, attempts to extrapolate the subsystem analysis trends into the future, both to look for potential problems and to allow planning of upcoming activities. Both of these processing functions require off-line software which uses spacecraft telemetry channels as input. The predictive function also needs for input the planned activities of both platform and platform subsystems. Figure 5.5 illustrates some of the telemetry analysis and performance prediction functions that can be performed by software. Some platform subsystems require only one of the two categories of software, while others need both. These are discussed in more detail below.

Telemetry-analysis software processes a set of past or current data from the platform subsystems to determine performance trends. In the thermal area, temperatures from on-board transducers are plotted to investigate thermal cycling of components that may occur due to any of several reasons: (1) the vehicle cycles in and out of sunlight when passing behind Earth or another planet; (2) the vehicle rotates so that thermally sensitive areas are alternately shaded and sunlit; and, (3) heat-producing components are turned on and off.

Thermal cycling is important because the expansion and contraction that results in certain materials can fatigue components and cause them to fail earlier than planned. High temperatures themselves can also cause parts, such as those in computer circuitry, to malfunction. Therefore significant effort is expended during typical mission operations to maintain platform component temperatures within specifications. Component temperatures must be monitored throughout the mission due to changing geometrical relationships, modifications in component usage, and gradual degradation in the protecting properties of thermal components such as radiators or blanket surfaces. Two of the most widely used software tools for thermal analysis of flight systems are TRASYS (Thermal Radiation Analysis System) and SINDA (Systems Improved Numerical Differencing Analyser), which have become NASA standard software packages (see Appendix 1).

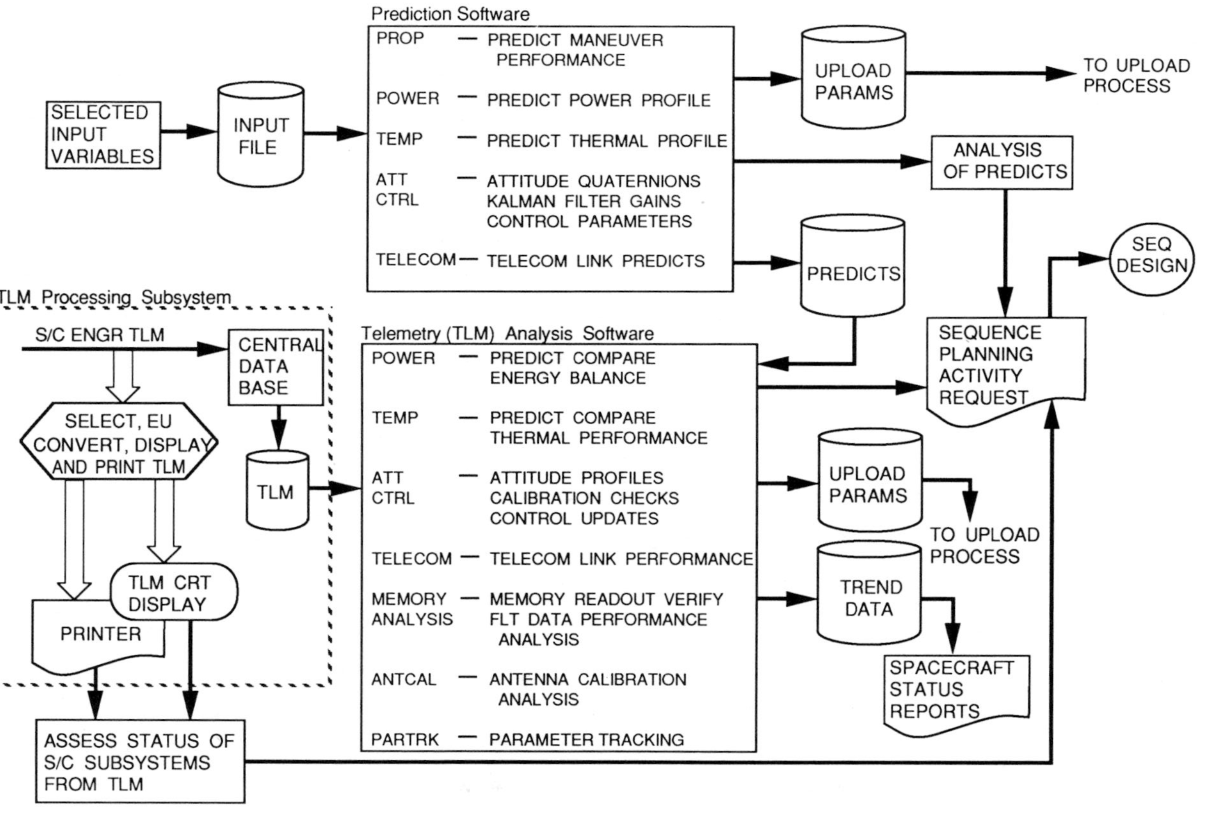

Figure 5.5 Engineering telemetry analysis flow.

The power subsystem is another case where analysis of telemetry is essential to proper functioning of an in-flight vehicle. In power systems using batteries, which are the most common type, it is important to maintain the energy balance between the batteries and the solar panels or RTGs. Here the main electrical bus current and voltage are carefully monitored during both charge and discharge cycles, and the depth-of-discharge of the battery is tracked. Since the supplied voltage may drop off if the depth-of-discharge is too great, (which is likely to be undesirable for some instruments or components), an adjustment must be made between the time spent charging the battery and the maximum sustained load on the system from all electrical components. Historically, batteries have been difficult to accurately model in software, primarily because each one responds individually and differently to a given load cycle. The best method appears to be to define a set of voltage-temperature curves that reflect the performance of the specific batteries to be flown and then adjust the curve parameters until flight performance is matched.

Attitude control is a subsystem area where there are generally multiple programs written to support telemetry analysis. Control of the platform's orientation in space is critical both for pointing an instrument to collect data and for pointing an antenna toward a ground station to play the collected data back to Earth. The absolute pointing positions, turn directions and turn rates all have to be carefully monitored for accuracy. Performance of gyroscopes including the motor current usage and drift rates are watched for any signs of abnormal functioning. Numerous parameters intricately involved with the technique used for maintaining attitude knowledge, whether it be star scans, Earth limb position measurements or other, require continued observation and adjustment as operational and geometric conditions change. On-board or ground algorithms for position and attitude numerical estimation may also need adjustment as a mission progresses by changing such parameters as mathematical filter gains. Since the variety of attitude control techniques and flight hardware components for remote-sensing missions is wide and usage is so dependent on the activity of a given mission, attitude control software is usually custom built for the specific application.

Telemetry analysis functions include verification of telecommunications link signal strengths to maintain a reliable signal between the platform radio subsystem and the ground station, and performing calibration consistency checks for in-flight functional calibrations. They also include analysis of regularly performed flight memory read-outs looking for bad memory locations, perhaps due to *single event upsets* (SEU), failures of memory cells due to high-energy cosmic rays which can either temporarily or permanently reset specific memory locations. The information obtained from the collected set of telemetry analyses is used to make decisions concerning how to operate various spacecraft components, especially those that might have a limited lifetime, and for possible alternative switches to back-up units. It also forms the principal input to periodic subsystem trend reports.

Performance prediction is accomplished whenever it is necessary to determine whether a desired sequence of upcoming activities is feasible. This provides a crucial link between the downlink process and the uplink process. For example, power prediction software would have as its input both the historical performance of the power subsystem in terms of energy provided under a given load, and the planned sequence of off–on cycles for all the power-using equipment on board. By extrapolating the performance of the power subsystem through the time period covered by the planned sequence, the sequence designer may find that for certain periods, there is not enough power to accomplish the desired activities. Adjustment of sequence activities back and forth in time might allow everything planned to be accomplished.

Similarly, the spacecraft thermal performance can be predicted, to see if any component thermal limits will be exceeded, by using a thermal computer model based on knowledge of how the spacecraft components dissipate, conduct, and radiate heat. Such a model requires a predicted attitude profile with respect to the Sun as an input to provide the externally introduced heat source. In cases where thermostatically controlled heaters cause a significant drain on the power system, the thermal predict program must be executed prior to the power prediction run in order to provide predictions of when the heaters will cycle on and off. It may be that the combination of one or more instruments and one or more heaters exceeds the capacity of the system. In addition, the thermal prediction program must account for the gradually degrading properties of the thermal components over the lifetime of the mission. This is accomplished by small adjustments to parameters in the thermal model that define the thermal characteristics (principally solar absorptivity) of the blankets, paint, reflective surfaces, and louvres.

Other subsystems also require a prediction capability in many missions. Spacecraft turn rate profiles, telecommunications link performance, and engine burn durations for manoeuvres are just some of the other performance parameters that can be predicted. Selection of guide stars, optional down link data rates, or engine nozzles (if more than one) are some of the decisions that could be made upon the outcome of a prediction. In many cases, a prediction program will be executed to provide predicted telemetry measurement time profiles, which can be plotted on a video display for subsequent overplotting of the actual telemetry values as they are received in real time. An example is shown in Figure 5.6, where real-time baseplate temperature telemetry overplots the prediction. If the predicts are accurate, such comparisons can provide an almost instant visual indication of an anomaly in progress.

5.7 User data processing

As for engineering data, the purpose of user data processing is to convert telemetry data into a form that can be analysed. User data go through a decommutation which is similar to that of engineering data with one excep-

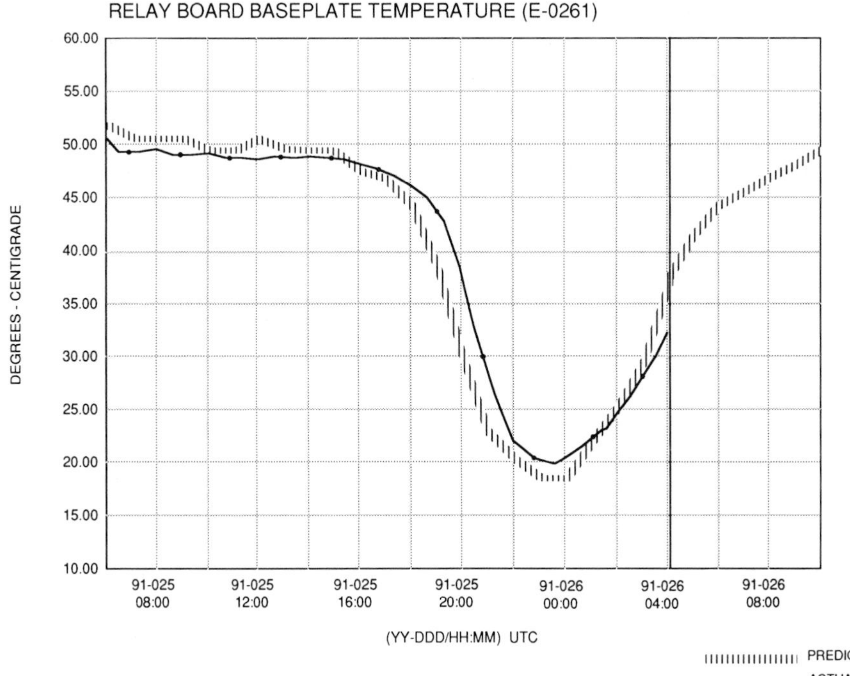

Figure 5.6 Example display of overplotting real-time telemetry on to a previously generated predict plot.

tion. As previously mentioned, most user data have a header at the beginning of each frame (after the frame identifier) which contains important ancillary data about the frame. Typical entries into the header are an echo of the command issued to take the data, acquisition time, frame number, line number (e.g., for images), and engineering data relevant to the instrument. The on-line telemetry process, generally a segment of the same process that handles engineering data, separates user data into a separate stream to create a data record called an *Experiment Data Record* (EDR). The EDR is, by definition, the primary archivable data record and should have the following properties:

(1) It should contain a minimum of redundant data (such as station handovers, where two antennas receive the same signal from the platform).
(2) It should contain all of the primary data or data from which all of the primary data can be recovered.
(3) It should either contain all ancillary data which would commonly be required for analysis or it should point to supplementary data records which do (see below).

Often, ground data such as ephemerides, orbit number, and other ancillary data are merged into the EDR. For some projects, the EDR represents the formal delivery of user data to the user.

The level of processing that EDRs require varies from almost none to very extensive, depending on the type and complexity of the sensor. Sometimes data processing consists only of performing decalibration and reformatting. Conversely, imaging data, which form the majority of data for many experiments, need extensive processing before they can be analysed. We will describe the EDR processing of imaging data in the following paragraphs as an example.

Vidicon and charge-coupled-detector imaging devices, imaging spectrometers and other image-forming sensors use the same basic method of acquiring data from a target, and the same basics in their data processing: the sensor dissects the target field into small elements and measures the quantity of interest of each (Huck *et al.*, 1975; Klaasen *et al.*, 1977). The resulting telemetry consists of data numbers (DN, or often dn) which represent brightness levels or other features of single-resolution elements of a picture, much the same as the elements of a newspaper picture. These elements, called *pixels*, each represent the brightness (or some other characteristic) of the corresponding element of the target. Together with the proper header, these DNs are delivered by the sensor to the platform data handling system, and downlinked to the ground.

During sensor testing the geometric relationship of each image pixel to the target pixel is determined. Using data from such tests, image pixels are assembled into a two-dimensional matrix placed in the same relationship to each other as the elements of the target. Then, using other sensor test data and perhaps calibration data taken in flight, the value of each DN is adjusted so that the resulting value bears a known relationship to the quantity being measured by the sensor (such as intensity). This action may involve the removal of a nonlinear response function, the subtraction of background noise, the removal of variations in sensitivity across the image, or other corrections. The matrix which results from these operations should be linearly related to the desired measurement, with the proportionality constant known to within understood limits. Image data are frequently subjected to one of several algorithms to enhance details or other salient features of the images, and multiple versions of each image then result (Castleman, 1979; Muller, 1988). Processed versions of EDR data, whether images or other sensor data, are known as *Team Data Records* (TDR).

For image data, the next step is to gather ancillary data and to assemble an image mask or collar. It is vital to a well-archived dataset that the essential ancillary data be present on any photographic print that is produced, to avoid the possibility of images that cannot be identified. While long used as a poor substitute for an image mask, ancillary data sheets have been discarded for that purpose as they are not automatically reproduced with the image itself. If more ancillary data exist than can reasonably be printed around an image, pointers such as file names are listed and the complete set of ancillary data is contained in the archive. Even in situations where photographic data products are not created, essential ancillary data must be preserved with the basic data. The image data are then merged into the mask, and the entire dataset is used to drive a film recorder which creates photographic products, generally on positive transparency stock. These master positives are used to create duplicates for distribution and archiving. Each science team member is typically given a set of negatives — along, of course, with the digital data.

The concept of adaptivity in missions will be discussed in Chapter 6. In the adaptive mission, the processing of user data carries urgency similar to, if not the same as, engineering data, so that decisions can be made concerning subsequent uplinks. Non-adaptive missions use off-line processors to process user data. This is not to say that there is no urgency attached to the processing of user data in the latter case, for it is only when the final products are available that the mission can say with certainty that the sensor is healthy and that the data have been taken correctly. For this reason, even non-adaptive missions have quality control procedures that quickly process at least some of the user data so that command errors can be identified and repaired. Even in situations where the identification of such problems would come too late to reacquire the same data, generic problems can be found and the quality of future data taken by the same system can be protected. Exhaustive data quality control procedures are employed in most missions to prevent problems from going undetected. Where the EDR processing is lengthy, temporary products are often created in a less rigorous way so that sensor health can be verified and problems identified quickly. The Magellan mission, where data from the DSN stations are sent to the control centre on computer tapes via mail systems, also sends back samples of the sensor data by electronic relay to create expedited versions of user data from each station to ensure that the sensor is healthy.

In the adaptive mission, however, user data processing is split into two streams. The EDR stream described above is carried out as previously described, but the telemetry processor also processes the sensor data for display in a way similar to the engineering display. Elementary forms of image-processing algorithms are implemented in the telemetry stream so that images can be displayed as they are received. The Viking first-order image processor for lander images allowed investigators to perform real-time searches for signs of life on Mars, and to also identify potential sampling sites as the data were received on Earth. Proposed Mars rover missions would require similar real-time processing to shorten the turn-around time for rover navigation.

5.8 Ancillary data processing

There is an advantage to placing all ancillary data necessary for analysis on the EDR, since then the analyst need go to only one place. Often, however, either because the resulting EDR would be too large or because ancillary data are updated after the EDR is created, *supplementary experiment data records* (SEDR) are created. These may contain position information not only for the spacecraft but also for moons and other bodies, instrument-pointing data, command data, etc. Science teams for past missions have defined five categories of data which should be contained on either the EDR or the SEDR:

(1) Ephemerides of the target body and other bodies of relevance such as the Sun.
(2) Ephemeris of the spacecraft.
(3) Instrument-pointing data.
(4) The command used to generate, store, and route the data.
(5) A recording of the actual (as opposed to intended) spacecraft activities that led to the creation of the data.

This set of categories forms the SPICE concept. The acronym represents the *S*atellite ephemeris, *P*robe ephemeris, *I*nstrument data, *C*ommand data, and *E*vent data.

5.9 Archiving

Starting with the preliminary design phase of a project, and continuing through to final close-out of the operations phase, it is necessary to keep in mind the group of people who will want to access the information gained by the project long after the mission is over, when those who conducted mission operations have gone on to other activities. Many who have found themselves members of this latter group know how frustrating it can be to have to reconstruct the processes that were followed, discover how to read the data records and what their contents mean, uncover project peculiarities, and so forth. Similarly, personnel leaving a project whose archive is less than complete will find that years later they are called on to explain what was done, how it was done, and many details they have either forgotten or wish that they had.

Projects have no stronger obligation than the one to their archive. In recent years this obligation has been strengthened for NASA missions with the formation of several permanent NASA archiving facilities, and with a formal agreement called the Project Data Management Plan between the project management and NASA. For other sponsors, this plan serves as an appropriate guide.

Two time periods need to be considered in the construction of an archive. First, it is important to archive a set of documents and data for the life of the project which allow project members to reconstruct what they have done, either for planning purposes or in the event of a failure. Such an archive can also be used to retain controlled copies of software and related documentation. Second, after the project has ceased to collect data the archive must contain adequate information to allow analysts to understand the user data and to interpret them in light of any relevant operational aspects of the platform. At the start of the latter period some of the data from the former period can be discarded. For NASA projects this is done as the archive is prepared for transfer to the long-term archive facility.

For both time periods, engineering as well as user data must be archived. During the mission, spacecraft engineering data are invaluable in tracing the cause of a failure. To this end, larger projects maintain on-line databases that can be called on to produce any telemetry from any past time period for comparison with more current data or to help troubleshoot a problem. A more limited set of engineering data should be carried over to the permanent archive to allow researchers to determine whether identified anomalies in the user data were caused by the target or by spacecraft anomalies.

General principles of archiving derived from experience with planetary projects can be summarized in a few rules which have proved useful:

(1) For all foreseeable reasonable research, and with only a reasonable effort, any future investigator should be able to find the data he requires together with enough ancillary information to carry on his work.
(2) If peculiarities are discovered in the data it should be possible to determine whether they were due to platform or payload malfunction or due to a characteristic of the target. Data-processing algorithms should be carefully documented for the same reason.
(3) A users' guide to the archive should be written, and it should be written at the close of the project so that processing artifacts and other oddities of the dataset can be described. This is the place to list changes made to the processing during the lifetime of the project, problems discovered in ancillary data, and other clues for future generations.

Table 5.1 is a list of data types to be considered for both end-of-mission and long-term archiving. This list was compiled from the Magellan archive contents, and it should be remembered that such a list is by nature highly specific to individual projects. In general an end-of-mission archive should contain copies of all actual unprocessed downlinked and uplinked telemetry, or as near to what was actually transmitted as the system will allow. After the project, a reasonable subset of these data should be preserved in the permanent archive. In addition, some textual history should be placed in the archive to preserve the intent of uplinked commands. Archival copies of all operational files such as decommutation and decalibration files need to be kept as well as operators' logbooks and other diaries. Source listings of all operational software, preferably in both human-readable and machine-readable form, must also be carefully retained.

Into the permanent archive go all user data, together with pertinent ancillary data and metadata. A necessary contribution of science teams during the development phase of a project is the definition of what ancillary data are to be placed in the permanent archive. Often forgotten in a permanent archive are detailed descriptions of how the data were processed; and, if processing

Table 5.1 Data types and retention periods for archiving

End-of-mission archive

- all raw downlink telemetry
- command development history
- decommutation files
- decalibration files
- controllers' logbooks
- sequence of events
- team status reports
- commanding and data processing software source listings

Long-term archive

- condensed downlink telemetry
- primary user data
- ancillary user data
- metadata concerning user data
- documentation of all operational versions of data processing software (including all algorithms)
- dates of software version changes
- command history
- commands as transmitted
- textual summary of command sent
- instrument calibration files
- instrument pre-flight test history
- spacecraft pre-flight test history

algorithms were changed during processing, descriptions of how and exactly when they were changed. It is prudent to include software processing version numbers and software installation dates in the data products themselves as well. The actual commands as transmitted and a text description of what they did should be included. The rationale for such information is that it is vital to be able to trace back the condition of the spacecraft when anomalies are seen in the user data, even years later, to ensure that a target characteristic can be separated from a platform or instrument problem.

Orthogonal to the above, there should be two distinct physical parts to an archive. Two copies of each data record submitted to the archive should be kept: the *working copy*, which is available for users to view, copy, and/or distribute according to the rules for that product; and the *archive copy* (or *deep archive copy*), which is kept separately and is not to be accessed at all unless the working copy is damaged or lost. Deep archive storage should ideally be under conditions prescribed by the manufacturer of whatever medium is employed for the data. Photographic negatives, magnetic tape, and discs all have their own preferences for long-term storage. NASA archive facilities have provisions for deep-archive storage. Projects, especially those whose operational periods extend for more than a few months, should mimic these conditions for their operational deep archives.

With the coming of more and more advanced database-management systems, the archiving process has become more sophisticated. Database-management systems employ complex cataloging systems, generally with some form of hierarchical structure to make it easier to find individual data records. It is now accepted practice to define such a structure as a project is designed, rather than to wait until delivery of an archive to its facility.

5.10 Summary

The downlink process has been described for both engineering data and for user data. The two data types are combined into one or sometimes two downlink streams on board the platform. Since bits in a downlink telemetry stream are a valuable resource, various data compression techniques are used to remove redundancy before transmission. This effort is balanced, however, by the need to protect signals from corruption by noise; some amount of redundancy is needed to provide tolerance to bit transmission errors.

As telemetry from a platform is received, other information about the incoming signal, such as Doppler frequency, is recorded to provide accurate estimates of the platform's trajectory. After receipt and digital conversion of the telemetry, engineering and user data are synchronized and separated. Engineering data, and sometimes limited amounts of user data, are automatically checked and displayed to flight controllers as soon as they arrive to identify possible problems. At a slower pace, the same data are analysed. Modelling software is used to predict future performance.

User data are processed into products which can be analysed, the details of which depend on the sensor which produced them. Properly processed and catalogued data products are then prepared for archive and, together with ancillary data describing them and how they were taken, are archived for future users.

Table 5.2 lists key terms from this chapter. Having now completed a description of the two basic processes necessary to the MOS, we will next see how these processes can be made resilient to things that go wrong, and how the MOS can adapt to discoveries made in the user data during the mission.

5.11 Exercises

(1) (a) Draw a detailed flowchart (at a level ready for translation into software) for a data-frame synchronizer. Its input should be a continuous bit stream, it should search through the stream looking for an n-bit synchronization pattern, and its output should be a continuous stream of data records, each m bits long. (b) Modify your flowchart to identify the same n-bit pattern even if the pattern contains one bit error. (c) Modify your flowchart so that,

Table 5.2 Key terms from Chapter 5

alarm-limit checking	metadata
alarming	original data record (ODR)
ancillary data	packetized telemetry
archive copy	parity
bit-error rate	performance prediction
caution and warning	pixel
commutated	science data
compressed	single-event upset
decalibration	subcommutation
decommutation	supplementary experiment data record (SEDR)
deep archive copy	synchronization word
encoded	team data record (TDR)
engineering data	telemetry analysis
flywheeling	tracking
frame	user data
hysteresis	variable packets
lossless	very-long baseline interferometry
lossy	working copy

once it has identified a few patterns, the algorithm will automatically increase its speed by flywheeling.

(2) In the sounding rocket example discussed in Section 5.3, if an anomaly caused the flight of the rocket to last for 37 minutes so that the clock rolled over, how might the ground system handle and process the last couple of minutes of telemetry?

(3) In Figure 5.2, the battery temperature sensor whose measurement is read out in byte 28 has a linear engineering unit range of −45 to +75° C. (a) What is the quantization accuracy of this measurement? (b) If telemetry arrives with a bit pattern of 10011100 for this measurement, what is its engineering value in ° C?

(4) If the actual nozzle temperature at one point during the flight of the sounding rocket in the example was 495° Centigrade, what would be the bit pattern returned in byte 22 of the engineering data frame if the sensor measures linearly over a range of zero to 600° C? What is the value of the error introduced by quantization in degrees?

(5) What is the bit pattern (12-bit mantissa plus sign bit and exponent) that would appear in bytes 17 and 18 for a sounding rocket yaw angle of 1° if the maximum range is ± 30°? The exponent contains a sign bit and two data bits and the calibration curve is linear. The sign bits are '1' for positive.

(6). A simple algorithm for a data compression scheme called run-length encoding operates as follows. If more than three bytes having the same value occur consecutively, the algorithm produces a code followed by the number of bytes having the value, followed by the value itself. Thus the data stream **7 8 8 8 8 8 8 8 8 8 9 13 13 15 15 15 15 15 15 15 15 3** would be transmitted as **7 255 0 9 8 0 255 9 13 13 255 0 8 15 0 255 3**. The code **255 0** marks the beginning of run-length encoding, and **0 255** indicates the return to nonencoded data (such a code would be chosen so as to be unlikely to appear in the natural data). Implement this algorithm in software and produce a graph of the number of bytes saved in the encoded data as a function of the number of identical bytes required. Evaluate this algorithm for different distributions of random and slowly-varying input data.

(7) Inject a bit error rate of 1:100 into the compressed data of exercise 6, decompress it with the errors present, and determine the number of resulting bit errors in the decompressed data by comparing with the input.

(8) Discuss the uses, advantages and disadvantages of each of the following display types: list displays, matrix displays and channel-versus-time plots.

(9) You are in charge of a power-limited spacecraft that is also very thermally sensitive. It has several remote-sensing instruments that must be carefully sequenced to avoid exceeding the power subsystem's ability to supply electrical power. It also contains several power-using, thermostatically-controlled heaters to keep certain components warm should the temperature drop below a given value. You have at your disposal four software programs as follows: SEQR is a program that builds a commandable sequence of events for operating the instruments in a specific order; POWR is a program that contains a model of the platform power subsystem and, given the right input, will predict the profile of power usage; TEMP is a program that models the thermal state of the entire vehicle and, given the correct input, will predict temperatures at given points; and PROF is a program that, given the sequence of events, will compute an attitude-time profile of the platform, including the direction of incoming solar heating. Determine what the input for each program must be; in what order the software must execute; if iterations are required to verify a given sequence; and in the event the power subsystem's capabilities are exceeded with a given instrument operation sequence, how must the task iterate to a final solution. Keep in mind that instruments drawing power also generate heat.

(10) For a free-flying platform with a hydrazine monopropellant propulsion system that can fire 33 pulses, each of 20 milliseconds duration, every second using thrusters or seven pulses of one-tenth second duration every second using engines, what functions should a performance prediction software program have to be able to predict the performance of an orbital trim manoeuvre? What telemetry measurements would be useful to predict?

(11) Discuss ways in which software might be made tolerant to single event upsets.

References

Castleman, K., 1979, *Digital Image Processing*, New York: Prentice-Hall.

Huck, F., McCall, H., Patterson, W. and Taylor, G., 1975, The Viking Mars lander camera, *Space Science Instrumentation*, **1**, 189–241.

Muller, J. (Ed.), 1988, *Digital Image Processing in Remote Sensing*, London: Taylor & Francis.

Klaasen, K., Thorpe, T. and Morabito, L., Inflight performance of the Viking visual imaging subsystem, *Applied Optics*, **16**, 3158–3170.

Kwok, R. and Johnson, W., 1989, Block adaptive quantization of Magellan SAR data, *IEEE Transactions on Geosciences and Remote Sensing*, **27**, 4.

Shannon, C. and Weaver, W., 1964, *The Mathematical Theory of Communication*, Urbana, IL: University of Illinois Press.

6

Anomalies, Contingency Plans, and Adaptivity

In past chapters we have briefly mentioned the concepts which form the title of this chapter. Now we will show how the basic mission operations system design developed thus far can be expanded to include the more pragmatic side of missions, where things do go wrong and where missions must strive to survive in spite of them. Experience has shown that although all missions must be preplanned, few if any missions are in fact conducted exactly as planned. Almost by definition, remote-sensing missions are complex; and that complexity dictates that surprises will and do occur. For the downlink elements of an MOS to be unprepared to identify anomalies, or for the mission planning or sequence generation or commanding elements to fail to be prepared to respond to them is to ignore the realities of operating a mission.

Any unexpected event whose import is great enough to have a significant effect on the operation of the mission, whether it occurs on the spacecraft or in the ground operations, is called an *anomaly*. The distinction will be made between an anomaly's *symptom*, the observable fact(s) by which the anomaly makes itself known, and the underlying cause, the originator of the symptom(s). A response or a potential response to an anomaly is called a *contingency response*, and a plan detailing such a response is a *contingency plan*. The objective of a contingency response is to restore the spacecraft to a safe and stable state from which planned operations can continue. They outline the intended mission response to foreseen potential anomalies.

After an anomaly has occurred, a contingency response invoked, and the system returned to a safe condition, performance of the platform or payload may once again be normal. Or, if the anomaly has left a permanent change, performance of the affected system may be degraded. If it is degraded, there may be a new set of limitations to the operation of the mission, and the mission goals may have to be adjusted appropriately through *alternate mission plans*. An alternate mission plan is distinguished from a contingency response in order to separate (in terms of both time and allocation of workforce) the often time-critical response to a safety-oriented situation from a more considered determination of how to reorganize mission goals. Thus, just as many anomalies which can be foreseen should have contingency responses prepared, those contingency responses which leave a system's performance degraded should have alternate mission plans prepared as well.

If no planning for anomalies has been done in advance, both contingency responses and alternate mission plans must be developed in near real time. Not only are valuable resources wasted, but the ability for rapid response to the anomaly is lost, which, in turn, could limit recovery options and even result in loss of the mission. However, if proper forethought has been given, both responses and plans can be agreed to by all parties in advance and implemented with little lost time. Even if the anomalies which actually occur are not the ones foreseen (and they usually are not), the experience gained from response and plan preparation gives considerable advantage to the mission operations teams in dealing with the problems actually encountered. As an illustration, the SIR-B payload's communication antenna was unable to properly track its TDRSS communication satellite — an anomaly. There was no contingency response prepared, for this anomaly had not been foreseen. The crew and ground teams spent valuable time devising a plan to freeze the antenna in one position so that the entire Shuttle could be rotated to track the TDRSS. As a result, SIR-B could not aim its antenna at targets while simultaneously relaying data to Earth, and an alternate plan was developed by SIR-B controllers which involved deletion of some targets and replanning others in order to maximize mission success. Most of the datatakes were salvaged, but much valuable time was spent in devising a response and an alternate plan.

When a non-nominal symptom is observed, the natural instinct is to try to understand its cause. At times, however, time does not permit this understanding and the symptom itself must be treated. Sometimes, the search for a cause can become secondary, especially when the system involved is removed from the investigators by large distances. The Magellan mission's attitude control computer infrequently begins executing an improper instruction (see Chapter 7), which may make itself known by loss of downlink and the subsequent invocation of fault-protection algorithms to regain contact with the ground, or by more subtle symptoms. A search for the cause of this anomaly was initiated after its first occurrence. That search was largely abandoned, although the cause was later discovered to be due to an undesirable flight software timing interaction. Meanwhile, some modifications to on-board software have been made to allow the spacecraft to be more tolerant to the condition. Ground procedures have also been established for more efficient recovery from the anomaly with minimal loss of science data. When a symptom demands urgent action, the search for a cause is not always the most effective course of action.

Figure 6.1 shows schematically how anomaly, contingency plan and alternate mission plan are related. When an anomalous symptom is observed (and perhaps while the search for a cause is underway), available preplanned

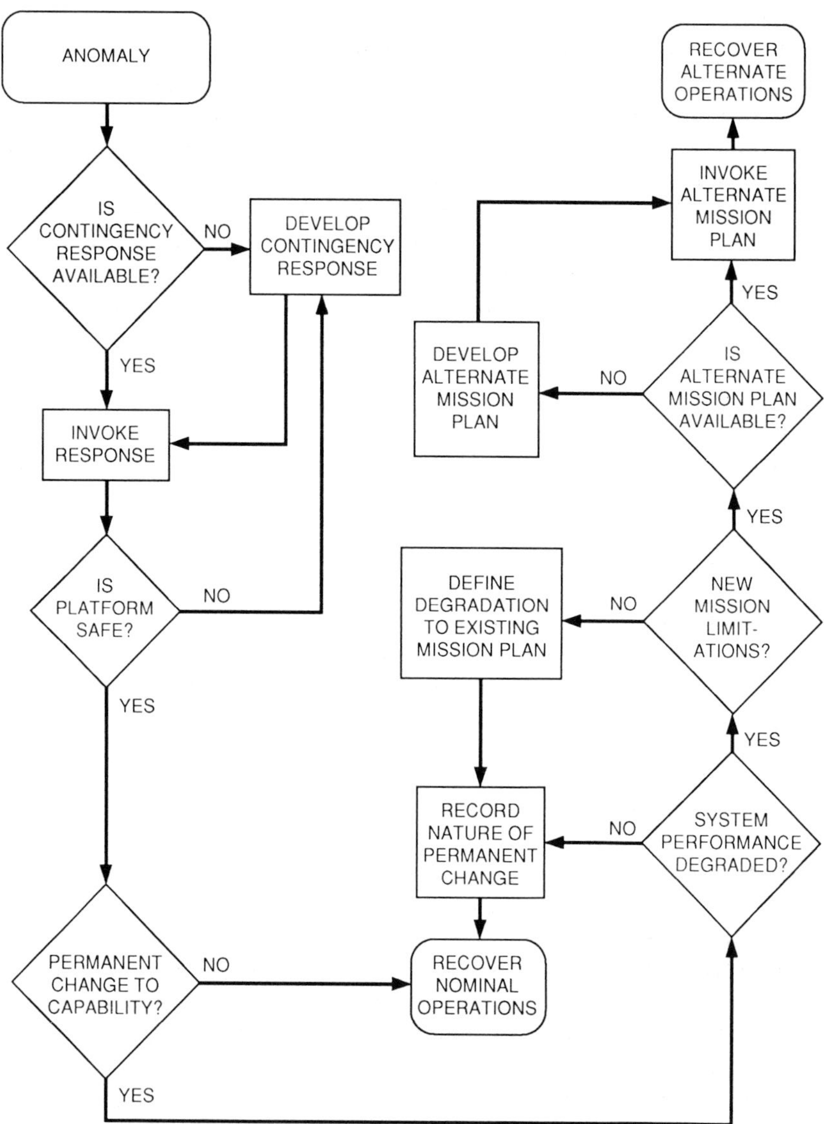

Figure 6.1 Decision flow chart showing paths from anomaly or symptom to invocation of contingency plan and alternate mission plan.

contingency responses are searched to identify one which is close enough to the observed situation to be useful. If found, it is invoked. If one cannot be found, and if there is no generic plan, a response is formulated and invoked. Then, if the response does not result in a safe platform, other responses are developed and invoked until the platform can be made safe. Next, the platform and its payload are analyzed to discover any residual changes to performance capability. If there are none, nominal operations can be re-established and continued. If capabilities have changed, have they resulted in a degradation of the performance? And if they have, can the original mission be carried out with those degradations? If the answer to the last question is no, it is only necessary to document the change and, if required, define the way in which the existing mission will be carried out in the presence of the degradation. If the mission must be altered, an alternate mission plan is required. If one exists, it is invoked; if not, one is developed. Finally, the altered mission is recovered and operations continued toward the achievement of its goals.

Adaptivity, the capability of a system to respond to the remotely-sensed data it has collected by modifying future data acquisition plans, requires an effort similar to anomaly response except that the goal is to gain additional science return. To add adaptivity to a mission can impressively increase the science return of many kinds of missions; but, as we will see, it is a complicated and expensive change to the design of the mission operations system.

6.1 Anomalies

Although in the popular sense anomaly response involves heroic action to save the platform from total loss, the more common anomalies are much more subtle. Most do not involve potential loss of the mission but instead require reaction in order to safeguard mission goals. It is not only the seriousness of an anomaly but also its probability and the capacity of the project to respond which determine how much advance planning must be done to prepare for its occurrence.

Preparation for anomalies begins by listing the most likely problems to occur as the mission proceeds. Table 6.1 shows a list of types of anomalies for a typical mission. Personnel absences, for example, are sure to occur during all but the briefest missions. They will often be due to sickness or holiday, but one must be prepared for personnel who leave or retire as well. Equally sure to occur are failures in ground computing equipment. For that reason all critical equipment should have back-ups, or at least a plan to work around the failed component or to operate without it. Mission planning minicomputers for the SIR-A flight were crippled when a floor polisher was plugged into the same power circuit which fed them. The resulting power fluctuations damaged parts on the computer power supply boards the night

Table 6.1 Types of anomalies which should be considered in mission operations

Ground systems

- personnel absences and changes
- ground hardware failures
- computer breakins (viruses, etc.)
- ground facility failures (bomb threats, earthquakes, etc.)

Flight systems

- flight hardware failures
- failed components
- failed memory cells (SEUs)
- memory parity errors
- flight software failures
- command errors
- launch failure
- abnormal manoeuvre performance

before launch. Had there not been extensive back-up planning tools prepared, the mission data return would have been seriously compromised.

Both because they are a long distance away and because they consist of special-purpose hardware, on-board anomalies deserve special attention when assembling an anomaly list. Flight memory chips are subject to damage from single event upsets, as mentioned in Chapter 5. Their frequency of occurrence varies with memory type and spacecraft environment, but for most missions their probability is high enough to warrant serious consideration. Although methods are employed in software to lessen the effect of SEUs, their effect can still be serious. Given the planned trajectory and lifetime of the platform, the probability of occurrence of an SEU can be calculated. If conditions warrant, SEUs must be anticipated with contingency plans and, in some cases, with software design which is tolerant to loss of single memory cells.

Also included in the anomaly list are significant deviations from nominal performance of critical subsystems. A good example is manoeuvre performance, which is never perfect but relatively easy to characterize using standard statistical techniques. Magellan's Venus Orbit Insertion engine performance must be beyond three standard deviations from expected before an emergency orbit trim manoeuvre would be required to avoid entering the venusian atmosphere, yet a contingency plan is available for such an occurrence.

6.2 Symptoms

Causes must not be confused with symptoms, either in the planning phase or during execution of a mission. For the purposes of this chapter, a *symptom* is defined as a directly observable non-nominal occurrence, such as a telemetry alarm. Many times an observed symptom may have more than one possible cause. A spacecraft manoeuvre which shows an excessive deviation from nominal performance is a symptom which could be the result of a propulsion system malfunction, a flight computer error or an incorrect manoeuvre design on the ground. The contingency plans for these three anomalies are quite different, and to implement the planned response to a stuck thruster valve when the anomaly was an incorrect manoeuvre design would bring disaster. For this reason, system engineers may devise symptom tracing charts which aid in tracing a set of symptoms to the anomaly which caused them and then to the proper contingency response. Figure 6.2 shows an example from SIR-A, which describes the action to be taken if the instrument is observed to be drawing too much power from the Shuttle. In the Shuttle operations these charts are referred to as malfunction procedures or 'malfs'. There they enable crew members to trace faults in complex systems about which they have only basic knowledge.

6.3 Contingency plans

Before contingency plans are developed for items on the anomaly list, each entry must be weighted by three factors: probability of occurrence, risk to mission success if the anomaly occurs, and ability of the mission to react to it. For Viking, failure of the lander to survive landing would have been a serious anomaly, but there would have been little useful reaction to that failure. Failure of Magellan to achieve Venus Orbit Insertion would have caused the on-board computer sequence to take several contingency steps in the minutes that followed the failure, but there was no useful contingency plan for those waiting on the ground, since the communications time is too long for them to effectively react. The mission would have become a total loss. The only anomalies which need contingency plans are those which have a reasonable probability of occurrence, a significant compromise to success, and a necessary and useful reaction.

Each entry on the list of most likely anomalies to occur which also poses high risk to mission success and has useful reactions should have a contingency plan associated with it. Of course, the realities of limited resources will determine how many plans can be developed, and to what level. In each case, the probability of the anomaly and impact magnitude must be traded off against the cost of preparation. A well-designed MOS will also include a generic contingency plan for what to do if no applicable plan exists. These plans should be reviewed and accepted by every team who is affected by the anomaly and every team who has a part in its implementation. Teams should consider whether the plans are appropriate and whether they could be implemented by that team in the presence of all other normal mission activity. Preparation of a reaction to and recovery from the anomaly may involve some work during the design phase (such as changing flight software to react

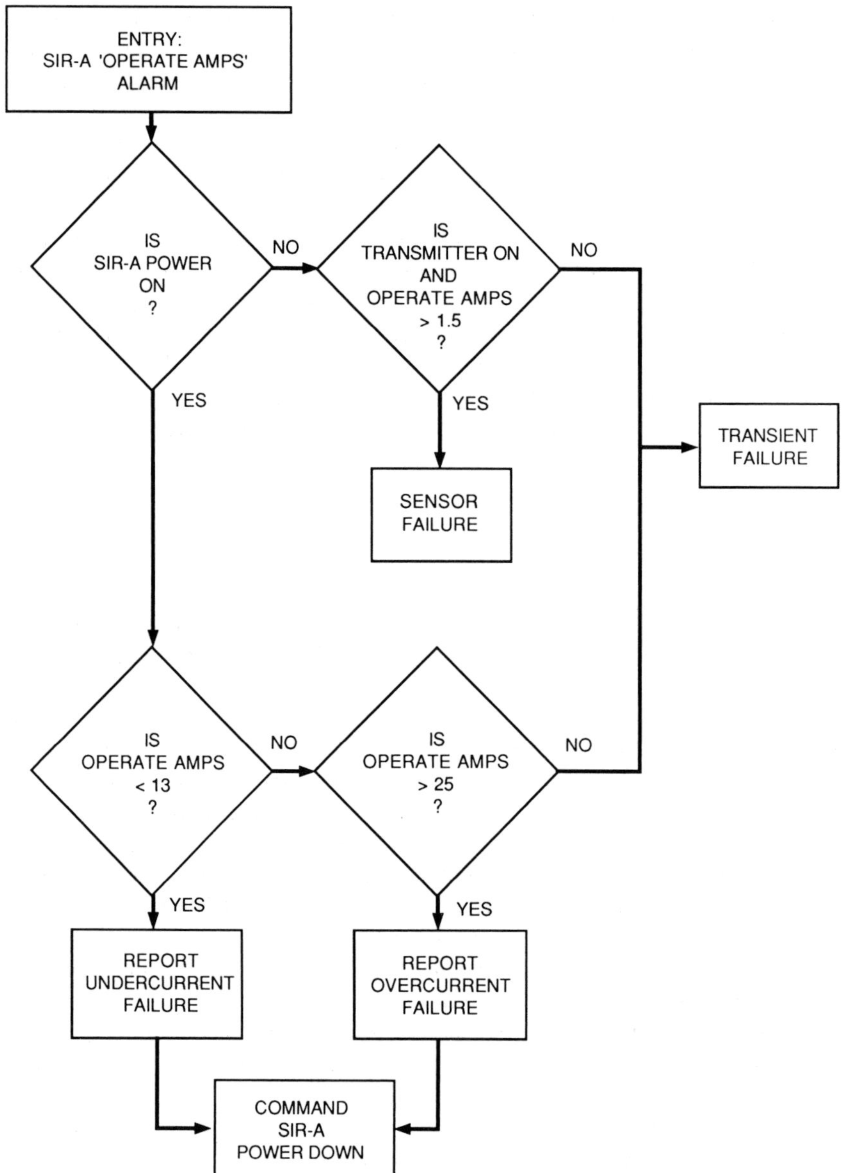

Figure 6.2 Symptom tracing chart from the SIR-A flight.

to the anomaly) or may imply some activity (such as a training exercise) to ensure readiness to react.

Contingency plans can be developed to several levels, depending on their likelihood, seriousness and on the time and resources available to the project for such development. Uplink teams may feel the necessity of preparing command sequences in advance which respond to certain anomalies, so that mission controllers can upload them immediately upon identification of the symptoms. For spacecraft anomalies which require uplink responses (and most will), four general levels of contingency preparedness exist. The lowest level, a plan which describes the response in text, may be sufficient for the less probable anomalies. Depending on the efficiency of the mission's command system, implementation of these plans may take from a few hours to a few weeks. More probable or more serious anomalies may require a second level of preparedness, a plan to identify what commands are to be used to respond; if the anomaly occurs, these might be developed into sequences, with the proper parameters chosen, within hours or days. Still more likely anomalies will require a third level, the development of command files with parameters already chosen. Such sequences might not provide any flexibility, but they are prepared completely and can be uplinked immediately. Finally, some sequences may be critical enough that they are run through the mission's spacecraft simulator in advance as a further time savings should they be required.

Figure 6.3 is a simplified list of Magellan contingency plans surrounding the Venus Orbit Insertion (VOI) phase of the mission. Orbit insertion, the deceleration of the spacecraft until it is brought into orbit around Venus, is followed by a period of careful tracking to ascertain the orbit parameters. Note that each item on the list is classified in terms of criticality, and each is assigned a development level. There is no direct correlation between the two; some anomalies capable of causing mission loss are only developed to the 'plan' level, while some less critical items have responses developed into sequences. In addition to criticality, both anomaly probability and resources required to develop the response play a role in this decision.

In order to aid in the decision as to which contingency plans are to be developed to which level, a concept called *fault tolerance* has been borrowed from the world of spaceflight hardware design. It is common in unmanned missions to decide beforehand that the operations must be prepared for any one of a set of failures to occur, but that consideration of two simultaneous but independent faults is not required. The system, hardware or operations design, is then said to be one-fault tolerant — it can tolerate one problem without diminished capacity, but not two at the same time. Critical parts of the Shuttle system, on the other hand, must be two-fault tolerant and must be operable even in the presence of two independent faults or anomalies. Life-support systems may be required to have even higher levels of fault tolerance. Based on its knowledge of acceptable risk, available workforce,

	ANOMALY	RESPONSE	CRITICALITY/RISK (1=highest)	DEVELOPMENT LEVEL
1	failed starcal prior to VOI	build option to re-perform any starcal if attitude error will be > 0.02 deg (n/a for VOI - 12 hours)	2/small attitude error at VOI	command
2	failure to achieve VOI attitude	recommand manoeuvre to VOI attitude	1/mission loss if burn attitude not achieved	sequence
3	small engine leak after burn	issue commands to fire pyro valves to isolate leaking engine string	2/loss of consumable at maximum rate of 0.5 kg/hr	sequence
4	pyro banks do not respond to "arm" command	command relays to alternate positions for either pyro bank	1(if both banks do not arm)/mission loss	test
5	failure of solar panels to achieve VOI position	recommand panel position out of either computer	2/undesirable panel dynamics and excessive contamination	sequence
6	single string computer error stops the VOI sequence in that string	build option to bring up dead computer and restart sequence. Plan assumes no hardware failure	2/higher risk of unsuccessful VOI on one string with additional failures	plan
7	earthquake knocks out JPL	store preliminary sequence and other non-sequence commands at the DSN stations	1/mission loss if JPL is unable to command spacecraft	plan
8	thruster overheat	provide capability to go to a thermally safe attitude post-VOI (especially for 2 days following VOI)	1/may exceed thruster temp sensor max and thruster temp limits	sequence
9	emergency orbit trim manoeuvre needed immediately after VOI	if periapsis < 150 km, s/c components will probably overheat	1/ degradation or loss of component in about 5 orbits	test

Figure 6.3 List of Magellan contingency plans for Venus Orbit Insertion (VOI).

and funding, a project can decide in advance what level of fault tolerance it is required to maintain, and plan for anomalies accordingly.

6.4 *Alternate mission plans*

The occurrence of an anomaly and the implementation of a contingency plan may leave the system as capable as it was originally — with no 'scars' from the event. For many potential anomalies there are remnant effects on the mission, such as diminished capabilities, which require either changes in the operations plan or a change to the goals of the mission. Each contingency plan should identify what differences are expected to remain in the system after the plan is implemented, and what reductions will exist in the capabilities of the system to accomplish its stated goals. A list of contingency plans which predict significant reductions in mission capability should be compiled, and for each entry either a brief statement of what changes would be made to the mission or a complete alternate mission plan should be prepared. An alternate mission plan must receive review by the implementors to ensure that it can be accomplished, and it must be approved by the users, so that in case it is needed there need be no debate about the propriety of its implementation.

6.5 *Anomaly reaction*

Anomalies are so mission-specific that it is difficult to generalize about reactions to them. Many issues need to be considered. Individual missions must carefully consider what procedures are to be established concerning anomaly response. The following issues for consideration are common, and we list them in order of their general priority, from highest to lowest. Individual missions, of course, must decide on their own priorities when writing contingency plans.

(1) *Determination of the appropriate contingency plan to be followed, or creation of one if none exists.* In many cases the specific anomaly that occurs will not be fully covered by an existing plan, and either a generic plan must be followed or a specific plan must be developed or modified in real time. Division of labour into those who will work the immediate problem and those who will plan the longer-term response may be appropriate. Similarly, the science impact of the anomaly should be determined and an alternate mission plan chosen if required.

(2) *Concerns for personnel/crew safety.* Situations which create or modify risk to human life must receive immediate attention. Such situations obviously occur in manned missions, but they also exist when considering response to earthquakes, fires and other natural disasters in ground-control centres.

(3) *Concerns for immediate spacecraft safety.* Is the platform or its payload on a

path which could lead to loss of the mission or other permanent harm? Should currently active datataking sequences be cancelled, or can they be allowed to continue? Should the spacecraft be commanded into a safe state, or is there risk involved in any commanding at all when the current status of the spacecraft is not understood? These and similar questions will affect any immediate actions to be taken to protect the spacecraft from further danger.

(4) *Collection of, or protection of, engineering or environmental data which may be useful to diagnose the problem.* When Magellan experienced a loss of telecommunications signal during a mapping sequence, a primary concern was implementing action to protect engineering data collected during the anomaly and recorded on its on-board tape recorder. Returning to the normal mapping sequence would have overwritten that data and destroyed potentially valuable clues to the cause of the anomaly.

(5) *Concerns for spacecraft's long-term safety due to its external environment.* When on-board systems are not behaving normally, ground controllers must consider what manual control may be appropriate to protect against environmental damage. Should the spacecraft be turned to shade itself from the thermal effects of the Sun while further anomaly resolution action is prepared?

(6) *Management of information about the anomaly.* Especially for larger projects, an important function during the progress of an anomaly is to manage the flow of information. It is vital for those trying to solve a problem to learn what others have discovered. A distinct but equally important issue is the dissemination of information to upper management, to the mission's sponsors, and to external entities such as the press. Rumours and hearsay are not the way for this to happen, yet unless someone is assigned to prepare and distribute regular reports these inherently less-accurate paths will develop.

(7) *Development of personnel shifting strategy.* Some anomalies are resolved in a matter of hours. Many, however, go on for days or longer. Positions which normally are staffed for normal working hours only may be required around the clock until at least the more immediate concerns are satisfied. While it is possible to stay awake and reasonably functional for many hours, it is seldom advisable to do so when there are other options available. When the determination is made that several days will be required to return an operational situation to normal, a shifting strategy should be devised which will allow necessary functions to continue. When an anomaly is discovered, the wisest course of action for management to follow may be to send half the workforce home to rest, so that they may populate a later shift. Otherwise they may well discover that, after 30 to 40 hr of continuous work by all available personnel, the anomaly is not over and everyone is too tired to continue.

(8) *Creation of a model of the anomaly that fits the symptoms from available data.* If an applicable symptom tracing chart has been drawn, this action will happen automatically. If not, some detective work will need to be done to isolate the most likely cause of the problem.

6.6 *Adaptivity*

Adaptivity is defined as the requirement for the mission to actively respond to characteristics or changes observed in the target by the sensor. In this way a single mission can accomplish the objectives of what otherwise might take two or more. An argument that has been advanced in favour of manned missions is that humans can react to what they see. Recent developments in the fields of artificial intelligence, imaging, and robotics have allowed unmanned missions to approach that capability without the associated risk and cost. Although the current state of robotic ability remains a poor second to human presence, the ability to adapt mission actions and even goals is a major step in that direction.

For example, the science return from a planetary orbital mission might be considerably enhanced by the ability to aim its spectrometer at areas which are identified from colour images as being of high scientific interest. General reconnaissance of the planet can then be combined with analysis of unexpected discovery. Had Voyager not had the ability to adapt its imaging sequences to the discovery of active volcanism on Io, another mission may well have been required to serve that purpose. With that capability, our understanding of the processes at work on that moon has now been advanced without further expense.

If a lander or rover mission is allowed to collect samples intelligently it will be able to much more comprehensively examine its surroundings than if it merely 'grabs' a sample with no forethought. Sample-gathering missions have been successfully run without being adaptive, but the science return is far greater if the platform is able to intelligently select its samples based on what has been learned from images. Several lunar sample return missions were conducted before astronauts aboard the Apollo missions carefully selected moon rocks, but the scientific return from the selected samples was far greater. A randomly-taken sample can give a completely wrong impression, so much so that sample return missions without the capability to examine many samples before choosing those to be returned are no longer viewed as scientifically worthwhile. If adaptivity is allowed, survey images can be taken to identify areas of interest, followed by higher resolution images to help set the sample acquisition strategy. Ranging may identify distances to prospective samples and enable collection of objects that are difficult to grasp.

For some missions, adaptivity is directly tied to the function of the mission. For example, a rover mission must be adaptive, as it is not safe to drive over terrain without first looking at it and reacting to what is seen. Many missions do not require adaptivity by their nature, or they may preclude adaptivity for cost reasons. A mapping mission such as Magellan has its nominal-mission goal well defined prior to the operational period, and implicit within its goal is that no matter what discoveries are made in the science data no change should be made to the mission objective of completing a global map of the surface of Venus. The addition of adaptivity can add significantly to the cost of a mission, and to some extent the Magellan decision was cost-driven. The Magellan design was preceded by VOIR (an acronym for Venus Orbiting Imaging Radar), also a mapping mission. But, without interrupting its mapping function, VOIR was to have the capability to take high-resolution images of features recognized in the mapping data. VOIR was cancelled for cost reasons, partly due to the inclusion of that capability. As currently designed, if Magellan were to stop mapping to re-image an especially interesting area, an outage in the global map (the primary mission goal) would result and coverage of possibly more interesting terrain would be lost. In their conceptual phase, missions must consciously decide whether or not to include the option of allowing adaptivity with due consideration of the resources which will be required.

It is clear that the scientific advantages of adaptivity are many in number. On the other side of this issue, however, the operational difficulties of implementing such a function are not to be underestimated when the mission operations system designers accept the job. To impose this requirement is to complicate the system considerably, and for that reason some missions have avoided it. As autonomy increases in robotic platforms and payloads, the ground-based effort required in the adaptivity loop will undoubtedly decrease. But the design must always take adaptivity into account, and the cost will still increase. Therefore this decision must always be made with foresight and consideration of the resource implications.

The complications are threefold. First, in nonadaptive missions that portion of the downlink process which is immediately downstream from mission control functions can be essentially removed from the pressures of real-time operations. This does not mean that there is no driver for the production of data products — such pressures come from the user as well as from financial considerations. But it does imply that those downlink functions can be spared the inevitable pressures on uplink and mission control to identify and solve problems on a normal operations schedule. In the adaptive mission the sensor data products are as time-driven as the engineering telemetry, and expedient operation of the data production function is as crucial as it is for the uplink function. Second, the data analysis process must also be carried out under a tightly controlled operations schedule. Data analysis, which generally is able to forsake expedience in favour of exactness, is not well suited to such an environment. For example, imaging sensors generally require some form of computer-intensive image processing in the data-processing stream in order to analyse the full resolution content of the data. When there is no schedule constraint, these processes can be carried out with ample operator interaction and on relatively inexpensive computer systems. It is not uncommon for some of these kinds of operations to require hours of computer time to complete. When a tight timeline requirement is imposed, however, significant changes to both hardware and software result. The Viking lander real-time image-processing system, although it employed only the most elementary algorithms, had to be coded at the assembly language level in order to meet the timing requirements of the mission. Finally, the up link subsystems must be designed to carry the extra load of adaptivity, involving additional interfaces, additional software testing and increased complexity. An adaptive mission requires a flexible and reactive up link process, the changing nature of which provides increased opportunity for commanding errors, unavoidably adding risk to the programme.

The absence of adaptivity, however, does not imply that a mission should not be able to respond to anomalies or be capable of redirecting its efforts should something go wrong. It may be a reasonable decision for a mission to decline the capability to adapt as we have defined it here. It is seldom reasonable to plan a mission so success-oriented that there is no capability to respond to anomalies, unless the sponsor is willing to accept very high risk of failure and to remain accepting of that risk when failure actually occurs. Missions have driven themselves to near failure trying to respond heroically to relatively minor problems which had been previously accepted as beyond their scope to repair.

Having drawn a clear distinction between anomaly response and adaptivity, it must be recognized that borderline situations do arise. Mariner 9 forms a good example: when that mission arrived at Mars, the planet was beginning what is now known to be a seasonally-occurring dust storm that can completely obscure the surface from orbital view (at least using its conventional imaging systems). The mission was fortunately able to survive this unexpected event and wait until the storm cleared months later before beginning its surveillance. It should be clear that a mission must be able to respond quickly to the unexpected when necessary to protect itself from failure, and that additional quick response capability to increase mission return is an expensive but desirable option. Proper distinction between necessary and desirable response is an important cost driver.

6.7 *Summary*

The more practical side to operating a remote-sensing mission has been discussed in this chapter. We have defined concepts which past missions have found useful in dealing with problems, and we have described generic principles that their MOSs have employed during problem resolution. Anomalies usually begin with the discovery of symptoms, which are sometimes treated as such, either while or in place of tracing them to their causes. Contingency plans are followed to resolve anomalies, and alternate mission plans are devised to continue to meet mission goals if the anomaly has left the mission less capable than it was.

Similar to anomaly response is the capability to respond to discoveries made about the target under investigation. An important difference between the two is that every mission must be able to respond to anomalies, but a

Table 6.2 Key terms from Chapter 6

adaptivity	information management
Alternate Mission Plan	Malfunction Procedure
anomaly	modelling
Contingency Plan	single-event upset
contingency response	symptom
fault tolerance	

mission may choose whether it can afford the ability to respond to scientific discoveries. The distinction between anomaly response and adaptivity is not always clear, and sometimes missions may have to decide on an individual basis whether a discovery can be left to a future mission. Table 6.2 lists important terms defined in this chapter.

In the following pages we will show how one particular mission has implemented the ideas developed in previous chapters, and how that mission made the distinction between anomaly response and adaptivity.

6.8 Exercises

(1) Using the definitions presented in this chapter, describe the equivalents of symptom, anomaly, contingency plan, and alternate mission plan that might be associated with (a) an automobile accident (b) winning the state lottery.

(2) Construct an anomaly list for a hypothetical Mars mission having an orbiter in a 6-hour orbit, a stationary lander and a rover with a 50-mile daily travel capability. List observable symptoms for each entry on your list.

(3) For the mission considered in Exercise 2, prioritize the considerations given in the 'anomaly reaction' section of this chapter. Then write a one- or two-sentence contingency plan for each item on your anomaly list and specify which plans should be developed into command files ready for use.

(4) From the list developed in Exercise 2, identify those anomalies which leave degraded performance, and for each briefly describe an alternate mission plan.

(5) A temporary but sudden and complete loss of downlink communications from a platform is a common anomaly. Such a symptom can be quite disheartening, since the tendency is often to fear the worst, but the real anomaly is usually much simpler. List as many anomalies as you can that would result in such a symptom.

(6) The planned Cassini mission includes a Saturn orbiter and a probe to be released into the atmosphere of its moon, Titan. Given that the probe can communicate with the orbiter, what are some of the ways the science data

return could be made adaptive during the probe release, atmospheric entry, and landed mission, especially since the atmospheric clouds are too dense to permit viewing the surface from orbit.

(7) Identify the parts of the mission operations organization presented in Chapter 3 (see Figure 3.3) whose involvement in real-time operations would sustain a significant increase if a previously defined non-adaptive mission were to be declared adaptive.

7

An Example — the Magellan Mission Operations System

In previous chapters we have described the general principles established by previous missions in the development and operation of their MOS. As with many attempts to generalize, it has been necessary to avoid some details in order to emphasize the commonality between these systems. The expense of doing so has been that many beneficial implementation details of individual missions have been left out of our discussion. To remedy this omission, at least in part, we present a more detailed view of a single MOS in this chapter.

Our choice for this example, the Magellan mission, is not to be regarded as necessarily typical but as a relatively modern implementation. Magellan has several characteristics that make its operations design different from most others, if not unique. First, Magellan was conceived as the first of NASA's low-cost missions and, according to its original design principles, was to accept somewhat higher risk in return for a reduction in cost, beginning with the design, build and test of the flight vehicle and continuing through the design and execution of mission operations. Second, operations for Magellan were, from the beginning, called on to take the first major advantage of a multi-mission operations system designed by JPL for NASA. Third, Magellan, as a long-term mapping mission, is not adaptive (as we have defined the word previously), and uses to maximum advantage its repetitive nature. Lastly, in a significant departure from previous JPL-led missions, Magellan's spacecraft engineers were to support real-time operations from a separate operations control centre 1500 km away in Denver, Colorado.

7.1 The Magellan mission

The Magellan mission has the goal of mapping the planet Venus using a single radar sensor which collects synthetic aperture radar (SAR) images (Elachi, 1988), and microwave altimetry and radiometry data. Another experiment uses the mass of the spacecraft and accurate tracking of the spacecraft's position from Earth to refine knowledge of the Venusian gravity field. The mission's scientific objectives are: (1) to provide a global characterization of observed landforms and tectonic features; (2) to distinguish and understand impact processes; (3) to define and explain erosional, depositional and chemical processes, and; (4) to model the interior density of Venus, especially to estimate the thickness of its lithosphere (Saunders *et al.*, 1990).

Early designs of the mission were named the Venus Orbiting Imaging Radar (VOIR). The VOIR design was rejected in 1982 for cost reasons, and a less costly version was developed under the constraint that it must be a low-cost, high-productivity mission, with a highly repetitive, non-adaptive operations philosophy. This philosophy developed into a mission design which necessitated a change from the desired circular orbit around Venus to an elliptical orbit that demanded less performance from the orbit insertion engine (allowing a solid rocket to be used). The primary effect was that science data would be collected at high altitudes over the planet's poles but at much lower altitudes near the equator. Less well foreseen was a dramatic increase in the complexity of both instrument and spacecraft control, which caused operation of the mission to become likewise complex.

The Magellan mission has been divided into three phases: prelaunch — the design, build and test phase — which lasted from the early 1980s until the May 1989 launch and included not only development of the spacecraft and instrument but also the design, build and testing of the mission operations system; cruise phase, lasting from launch until Venus Orbit Insertion in August 1990; and orbital operations phase, which included a short check-out period following orbit insertion followed by radar mapping of the planet's surface which continues at this writing. The post-launch phases are illustrated in Figure 7.1.

A short introduction to the spacecraft may help the reader to fully understand the mission. A photograph of the Magellan spacecraft was shown in Figure 1.2. This figure will be referenced here, as components will be described starting with the antennas at the top of the picture and working toward the bottom of the vehicle. The High Gain Antenna (HGA) is the large dish-shaped antenna at the top of the vehicle. It is used both for communications with Earth and for transmitting and receiving reflected radar signals from the surface of the planet. The HGA's secondary reflector is seen sitting on top of the feed support structure. On the very top is the low-gain omni-directional command receipt antenna affixed to the back of the reflector. This antenna is the emergency commanding antenna. To the left of the HGA is the altimeter antenna, a long narrow structure fixed at a permanent angle to the HGA boresight.

Below the HGA is the forward equipment module (FEM), a rectangular structure that is approximately half occupied by the electronics for the radar and altimeter. The star scanner, an optical telescope with an electronic detector at the focal plane, looks out of the side of the FEM facing the viewer. On the opposite side of the FEM from the star scanner, and out of view in the photograph, is the medium-gain antenna (MGA), which was used for

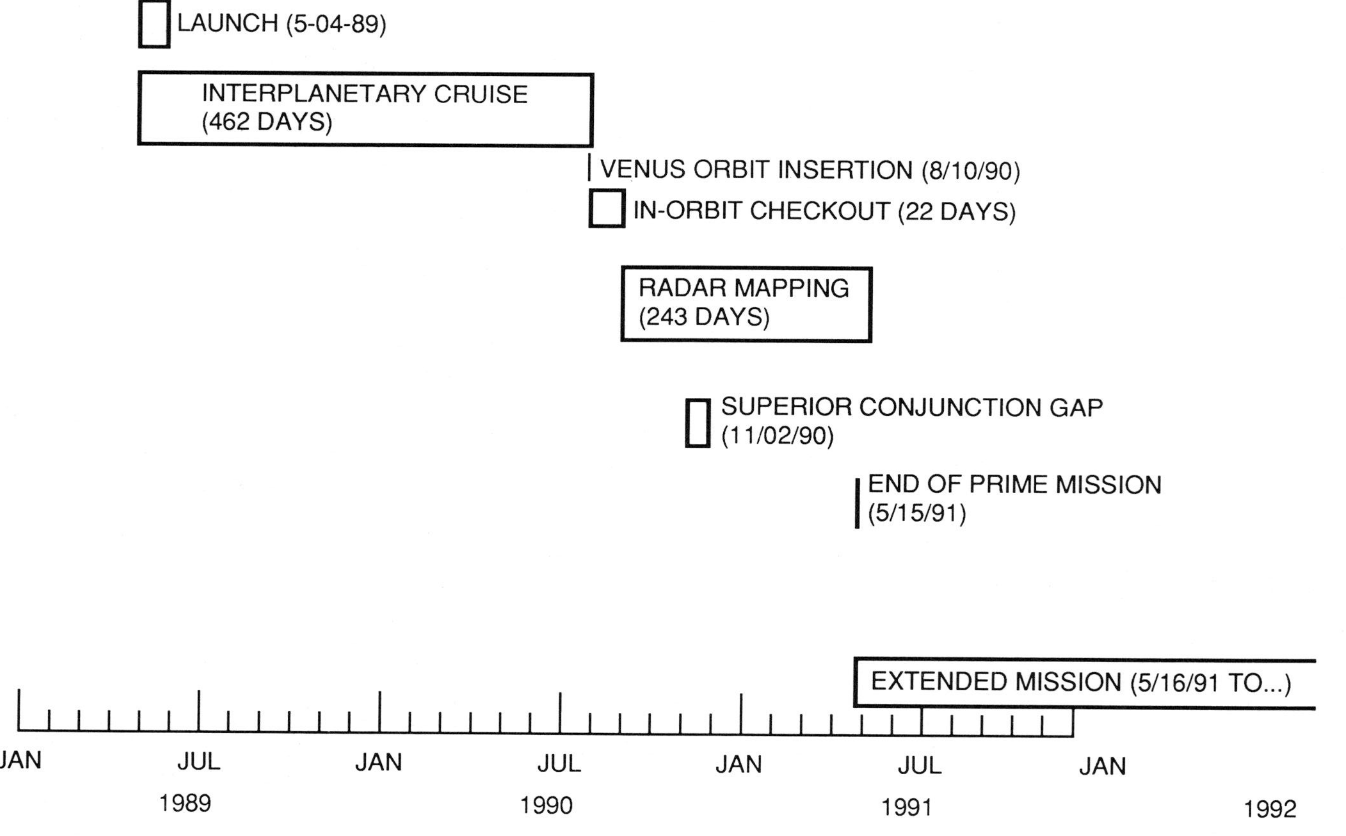

Figure 7.1 Magellan mission timeline.

communications during cruise and is available for emergency downlink. The remainder of the components that reside in the FEM are the radio electronics, two batteries, the star scanner electronics, the gyroscopes and their associated attitude reference unit electronics and three momentum wheels for turning the spacecraft. At the base of the FEM, on either side, are the arms for the solar panels, with the articulation devices at each end. The two solar panels are folded downward in this photograph. They were deployed shortly after launch and have remained deployed since.

Below the FEM is the 10-sided structural bus, a spare from the Voyager programme. The various bays of the bus contain the controls for the solar array drive motors, the two attitude control computers, the two command and data system computers, two tape recorders, four bulk utility memories, the power switching unit and the power distribution unit. In the middle of the hollow bus ring, and out of sight in this photograph behind thermal blankets, is the hydrazine propellant tank. The engines and thrusters of the propulsion module itself can be seen at the ends of the struts below the bus. Each of the four rocket engine modules contains two 440 N engines, a 22 N thruster and three 0.9 N thrusters, as shown in Figure 1.6. These were used for mid-course trajectory corrections, orbital trim manoeuvres and to desaturate the momentum wheels. At the bottom of the picture is the solid rocket motor that put the spacecraft safely and accurately into orbit around Venus and was then jettisoned.

To understand why the Magellan mission operations system operates the way it does, the details of the mapping phase must be made clear. From an operations standpoint, Magellan repeats the same pattern every orbit as Venus rotates underneath the inertially-fixed, 3.26 hr elliptical polar orbit. Figure 7.2 shows the sequence of events. As the spacecraft approaches the North pole, it turns its HGA toward the pole, pointing at a look angle of about 15° with respect to *nadir* (i.e., straight down toward the surface directly under the spacecraft) and perpendicular to the direction of motion of the ground track, and begins to take radar data, storing it on flight tape recorders. As it approaches periapsis, the closest approach to the planet's surface, it gradually rolls the high-gain antenna away from nadir to increase the imaging incidence angle (more will be said about this in a subsequent section), then pulls it back toward nadir as it nears the South pole. At the conclusion of its mapping pass, the spacecraft turns the high-gain antenna to point to Earth. After a short period to allow the DSN antenna on the ground to acquire telemetry lock, the on-board tape recorders begin to play back the recorded science data. Near apoapsis, the playback is temporarily halted as the spacecraft updates its attitude knowledge by scanning a preselected pair of reference stars, despins its momentum wheels if necessary, and resumes playback. It concludes playback in time to reorient itself over the North pole for the next mapping pass. Note that during the mapping pass the spacecraft is completely out of contact with Earth.

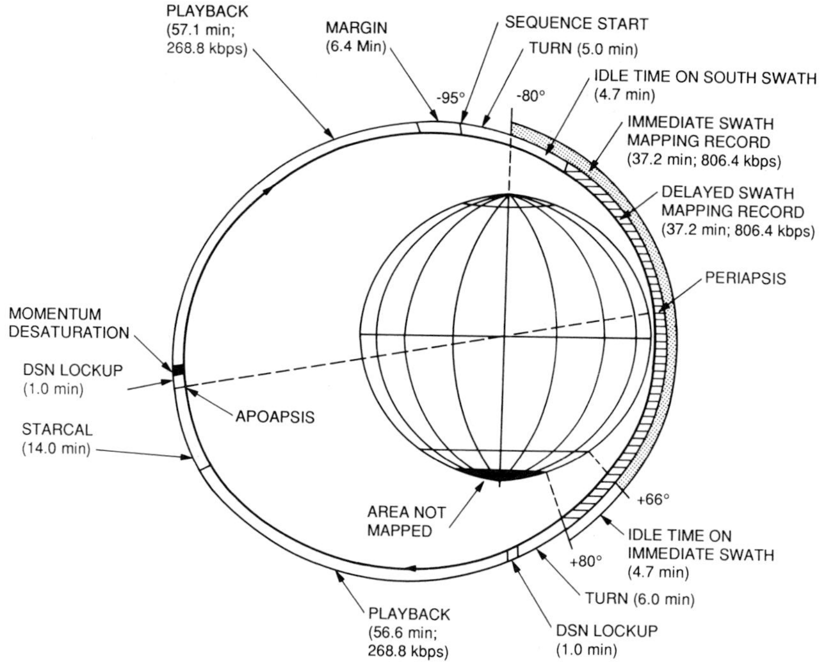

Figure 7.2 Magellan mapping orbit.

7.2 The radar sensor

The functions of SAR, altimeter and radiometer are all combined into a single instrument which is called the radar sensor. This equipment performs the precise timing required to send out a series of high-power pulses of microwave energy through one of two separate antennas and then to receive the echoes of those pulses from the planet's surface at times between pulses when radiation is not being sent out. The burst of pulses required to do this is illustrated in the timeline of Figure 7.3. This burst is repeated throughout each mapping pass, covering on alternating passes the area from Venus' north pole to about 50° South latitude or from about 55° North latitude to 75° South latitude.

Figure 7.4 shows conceptually the operation of the SAR and altimeter portions of the sensor. The attitude of the spacecraft is maintained so that the HGA points to the side while the altimeter antenna points directly toward the surface. The direction of motion of the spacecraft is perpendicular to the plane of the diagram. Because the SAR pulses are sent out at an angle, the portion of each pulse that is reflected from surface elements nearest the spacecraft return before those farther out. The returned pulse can be divided into segments corresponding to different surface elements, and an image-like

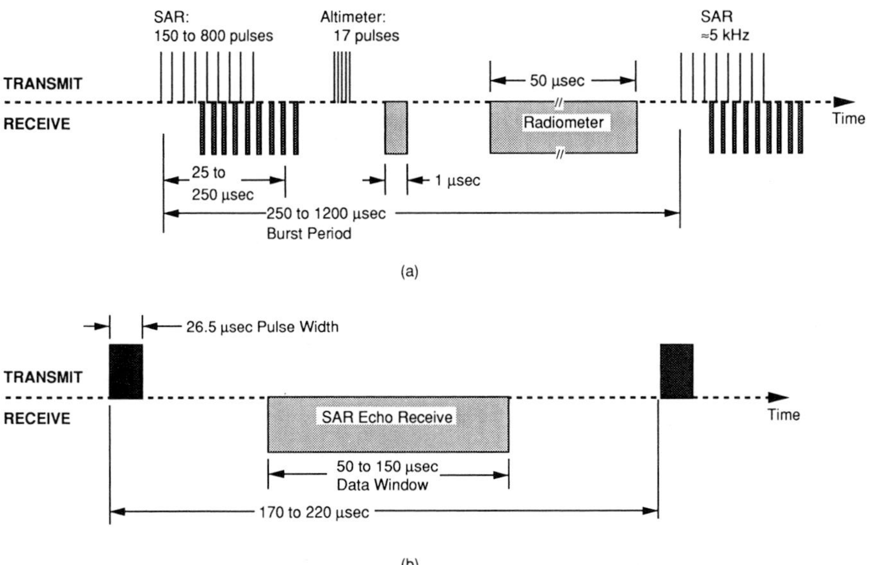

Figure 7.3 Radar sensor timing diagram: (a) shows timing of the entire burst; (b) shows an individual pulse and the following receive window.

array can be formed of these segments because of this time separation. The second dimension of the image is formed by recording the *Doppler shift* of each segment and resorting the array. Meanwhile, the altimeter pulses are also timed, and the time required for each pulse to be reflected and returned determines the distance from the spacecraft to the surface element directly below. This information is combined with knowledge of the spacecraft orbit to produce a topographic map of the surface.

Each burst begins with a series of several hundred pulses from the high-gain antenna, which is pointed at a side look angle to the nadir direction. Pauses are inserted between pulses to allow echoes from the initial pulses to be received, as shown. Next, 17 pulses are sent from the altimetry antenna, which is pointed directly toward nadir. Transmission is then stopped for a period adequate for receipt of all altimetry pulses, and then the receiver listens through the HGA once again for the natural microwave emissions of the planet. Note that the placement and timing of events within this burst depend critically on the time required for a pulse to travel to the surface and back to the antenna, which is a function of the slant range from the antenna to the target point on the ground at which it is aimed, which in turn depends on the altitude of the spacecraft above the surface and the side look angle at which the antenna is pointed. Thus the parameters which control the burst are highly sensitive to both attitude and position of the spacecraft. Given the rapid change in spacecraft altitude above the surface which occurs during

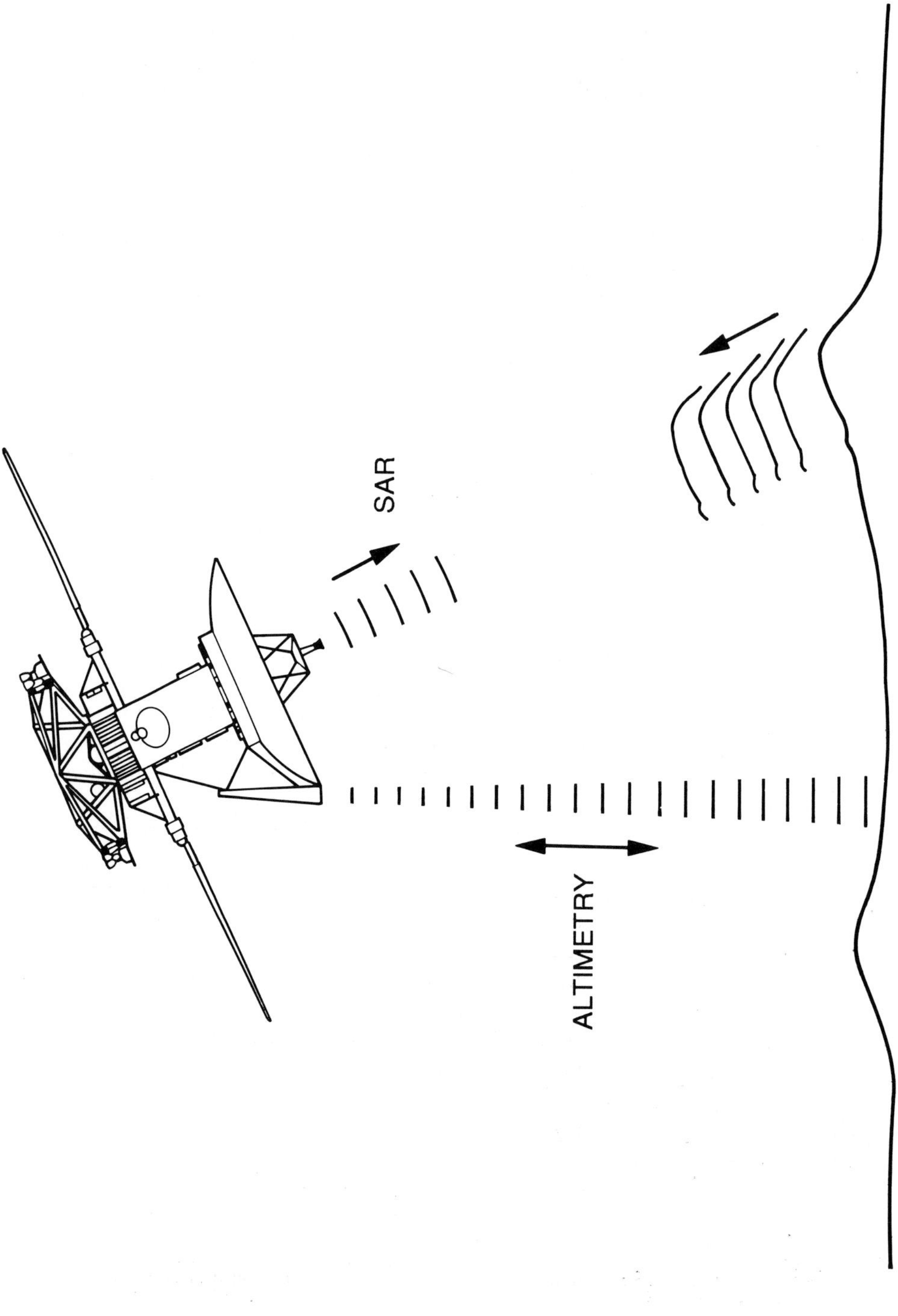

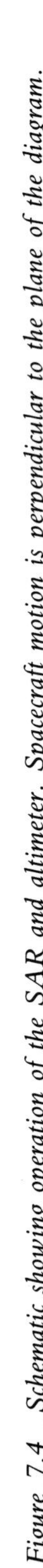

Figure 7.4 Schematic showing operation of the SAR and altimeter. Spacecraft motion is perpendicular to the plane of the diagram.

each mapping pass, it is clear that many changes must be made to these parameters, and that the values given to them require very accurate knowledge of both spacecraft position and attitude as a function of time.

The ability to understand objects seen in images, and particularly SAR images, is strongly dependent on the angle at which the object is viewed. With SAR, better resolution is obtainable in the cross-track direction (i.e., the direction perpendicular to the spacecraft's ground track) as the SAR looks farther away from nadir. Thus, to obtain best resolution, the SAR needs to look as far to the side as possible. Unfortunately this geometry also increases the distance from the antenna to the surface, and as that distance increases the signal strength dissipates and the images become more noisy. Since for Magellan's elliptical orbit that distance also increases near the poles, the side-looking angle can be greater near periapsis than at the poles. In order to take advantage of that possibility, Magellan rotates slowly about the spacecraft's velocity vector as it acquires radar data, from an off-nadir angle of about 15° at the pole to about 45° at periapsis. In this way the changes in distance from antenna to surface are minimized at the expense of a more complex spacecraft operation. The angle between altimeter antenna boresight and HGA boresight plus the altimeter beam width of ± 15° in the cross-track direction permits this rotation while allowing continued collection of altimetry data.

7.3 Overview of the Magellan mission operations system

Magellan's uplink process satisfies mission objectives by developing and uploading sequences of platform and payload activities that respond to requests for user or engineering data, interspersed with supporting ground events. The process is divided into a number of tasks which define the sequence design steps. These are defined as advanced sequence planning, sequence planning, sequence generation, and command processing. In advance sequence planning, a time-ordered list of activities is developed based on a mission plan, which is composed of requirements and constraints at a high level. A sequence is then formed which flows from task to task where it is repetitively refined. In the command processing task, a command file of intended activities is radiated to the spacecraft.

Figure 7.5 shows how the uplink and downlink processes work together. The Mission Plan is periodically updated, generally once per mission phase. Thus, a Mission Plan is written for cruise, updated for the first 'cycle' (or 243-day Venus rotation), updated again for the second cycle, and so on. Each version of the Mission Plan is used to generate advanced sequence planning. An advanced sequence plan is combined with ground events such as ground antenna coverage requirements and sent to the first phase of sequence planning. In sequence planning, the preliminary sequence of commands is produced. Some iteration is commonly required between the sequence designers

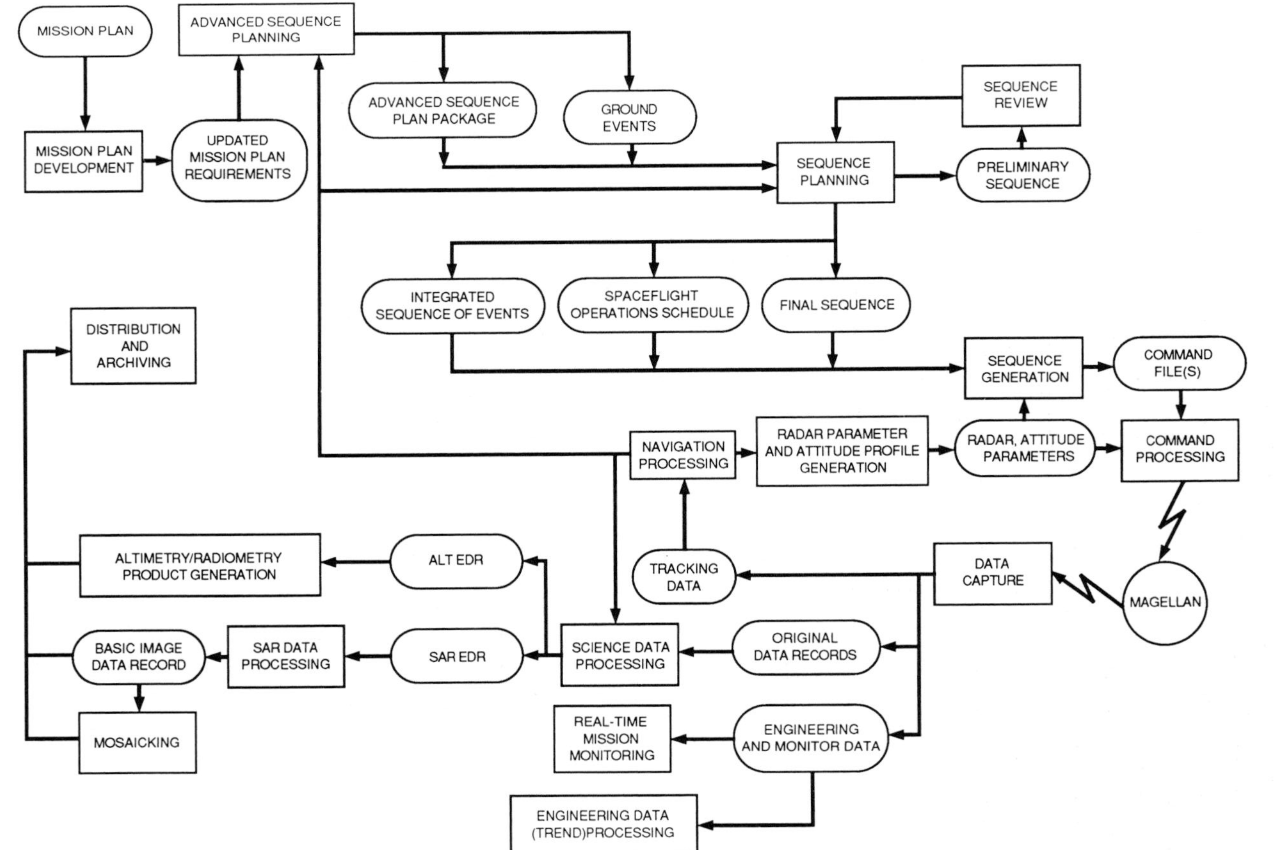

Figure 7.5 Overview of the Magellan mission operations system (mapping phase). Tasks are shown as boxes with squared corners, while products are shown with rounded corners.

and the sequence builders as a thorough review of the preliminary sequence is conducted and the details in the sequence are filled in. After such iteration a final sequence design is produced, along with two levels of printed schedules, whose detailed description is left for a later section.

The final sequence is transformed into a command file and, together with the schedules, is transferred to the DSN. Even though Magellan is not by nature adaptive to science discoveries, data analysis results are considered on a long-term basis, and of course anomalies causing gaps in the mapping data can affect plans for following uplinks. Engineering data analysis results are also considered at all levels of sequence design, since non-nominal conditions on the spacecraft may have an influence on future commanding.

Magellan's downlink process is complicated by its high science data rate requirement. Radar data are played from the on-board tape recorder at a rate of 268.8 kbits/s, which is higher than the capability of the DSN to transmit data from their station complexes to the network control centre at JPL. Therefore most of Magellan's science data are transmitted to JPL by computer tapes which are shipped from the individual DSN complexes. In order to verify that the sensor is healthy and is being properly commanded, a small percentage is selected at the individual stations and transferred by electronic means at a lower rate. This creates two situations which complicate the mission operation. First, a significant percentage of the radar data must be handled twice — once for health analysis and again to process the science data. Second, much of the mapping data arrival is delayed by up to several weeks. As tapes arrive from California, Spain and Australia, the data are received in an unpredictable order and must be staged until mosaics can be assembled in the order that the individual mapping strips were taken.

Independent of the science high rate operation is a lower rate channel, at 1.2 kbits/s, of engineering telemetry which is transferred to JPL in real time, along with ground processing monitor data, to monitor the health of all spacecraft subsystems, especially those critical to current operations. These data are decommutated, decalibrated, and alarm limit checked as they are displayed on screens in front of spacecraft controllers who monitor the health of the spacecraft and its instrumentation 24 hrs each day.

7.4 *Project management and organization*

The Magellan programme is managed by NASA, under the direction of a Programme Manager and a Programme Scientist. JPL designed, built, and now operates the Magellan project for NASA together with its prime contractors, Martin Marietta Astronautics Group and Hughes Aircraft Company. JPL's effort is led by a Project Manager (PM). Figure 7.6 presents the project organization for mission operations.

In addition to the usual financial and other administrative assistance, the Project Managers' staff consists of a Project Scientist, assisted by a subset of

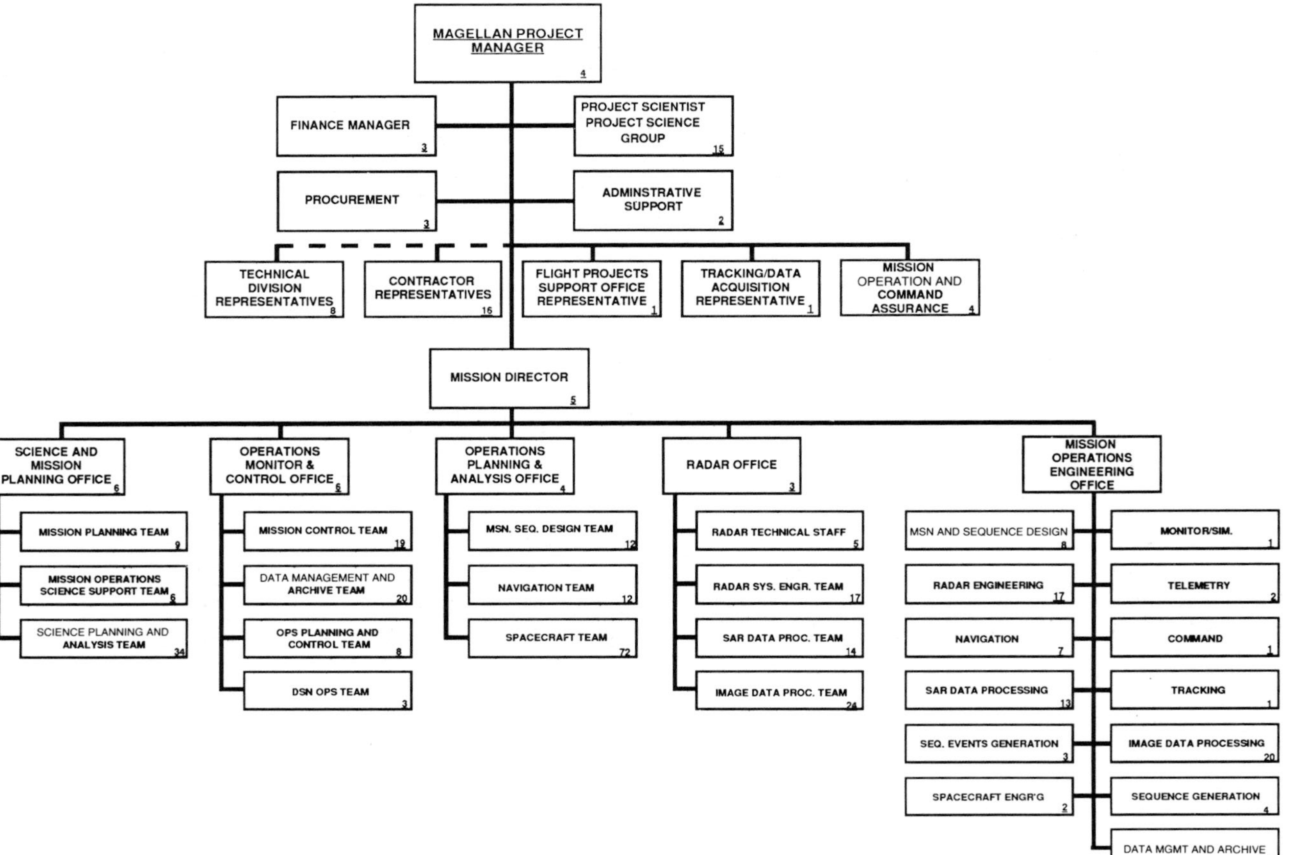

Figure 7.6 Magellan project organization chart. Numbers in lower right corners refer to number of personnel.

Magellan's scientists known as the Project Science Group (PSG). The Project Scientist and the PSG formulate the project's science policy and advise the PM on issues involving modification to or clarification of that policy. The PSG has two working groups to formulate such advice, the Data Products Working Group and the Mission Operations and Sequence Planning Working Group. Their advice is passed on to the PSG in the form of requests for action. If the PSG concurs with these requests, the Project Scientist carries the requests to the PM. Other staff assistants to the PM are representatives of: (1) the prime contractors; (2) the technical divisions of JPL who support the project; and (3) the DSN management (an office called Tracking and Data Acquisition). Finally, an office called Mission Operations and Command Assurance provides quality assurance for the MOS.

The Mission Director (MD) reports to the PM directly. His responsibility is for overall operation of both uplink and downlink processes. Although the MD position is not normally staffed more than one shift, five days each week, his office must approve any command activity. Thus, two Deputy Mission Directors provide round-the-clock coverage of the MD position during periods of high activity or anomaly reaction. There is also an Assistant Mission Director, who manages the day-to-day activities and problems in the downlink process.

Under the MD are five offices, within which lie the operations teams. The Science and Mission Planning Office (SAMPO) houses the Science Planning and Analysis Team, which contains the two Magellan science groups, the Radar Investigation Group and the Gravity Investigation Group. Subsets of these two groups are independently part of the PSG and its working groups. SAMPO also contains the Mission Planning Team, which not only prepares the Mission Plan and Advance Sequence Planning Package as described below but also writes mission-level Contingency Plans. Finally, the Mission Operations Science Support Team has two Science Coordinators, one for uplink and one for downlink. These two positions act as interfaces for the two corresponding PSG working groups.

The Operations Monitor and Control Office manages the personnel directly involved with data manipulation and monitoring. This includes: (1) the Mission Control Team, charged with 24-hour a day monitoring of spacecraft telemetry and with generation of the integrated sequence of events product for flight team distribution; (2) the Data Management and Archive Team, responsible for handling data products and for maintaining the project data archives; (3) the DSN Ops Team, which interfaces with the DSN for scheduling Magellan activity; and (4) the Operations Planning and Control Team, which is responsible for operating the Magellan portion of the JPL multi-mission operations facility.

A third office, the Operations Planning and Analysis Office, (OPAO) has responsibility for non-real-time functions. Included in this office are:

(1) The Mission Sequence and Design Team, responsible for taking inputs for the Advanced Sequence Planning Package and developing the preliminary

sequence of spacecraft and instrument events to accomplish the requested activities;

(2) The Navigation Team, whose job is to process the spacecraft tracking data into orbit determination products to maintain precise knowledge of where the vehicle is at all times and to support manoeuver planning if changes in the trajectory are needed;

(3) The Spacecraft Team, which is responsible for the health and safety of the in-flight vehicle, from the standpoints of command generation and engineering telemetry trend analysis.

Magellan organization reflects a basic difference from most planetary missions in that the spacecraft contains only one instrument. As a result, the radar sensor operation, data processing and image mosaicking operations are managed out of a dedicated office called the Radar Office. The Radar Office contains four teams, each of which directs one phase of the radar operation. The Radar System Engineering Team develops the radar control parameters and analyses (in non-real time) the radar engineering and science data to monitor sensor health. The SAR Data-Processing Team accepts sensor and ancillary data and produces the individual strips of SAR data. The Image Data-Processing Team receives the SAR image strips and creates mosaics and other products for use by the science teams. Also in the Radar Office is a Radar Sensor Chief Engineer and his staff, which compose the Radar Technical Staff.

In a separate office, the Mission Operations Engineering Office, lie the developers and sustainers of the various software and hardware subsystems that support the corresponding operations teams. Each team is represented in this office by one or more cognizant engineers who are responsible for operation of the team's subsystems. When a team recognizes a need for a change to a subsystem, and after the MD has approved such a change, these engineers are responsible for implementing change to the subsystem, and for the required testing and integration of the newly-delivered subsystem into the overall MOS.

7.5 *The uplink process*

Most of Magellan's planned commanding is performed through its standard command process. This activity is designed to reliably create uploads that perform the same basic behaviour repeatedly, with only minor changes to parameters and command details occurring with each implementation, supporting the repetitive mapping process. Two other activities are also defined that can command the spacecraft in a less routine manner.

The Magellan mission plan document serves as a source of requirements and constraints, anticipated DSN support requirements, and the latest engineering analysis of spacecraft capabilities. The Mission Plan is used to begin sequence planning, the first step in the uplink process which will create sequences of commands to control Magellan for one week each. The first version of an activity plan, created directly from the mission plan, is contained in the advanced sequence planning package (ASPP). In this package, a time-ordered listing of ground and spacecraft events is presented to the flight teams for review.

Five of the nine teams involved in the standard sequence uplink process initiate sequence planning. The Mission Planning Team coordinates and provides sequence-specific, mission-level requirements and constraints based on the Mission Plan and inputs from other MOS teams. In particular, the Mission Operations Science Support Team, led by the Magellan Experiment Representative, negotiates compromises between conflicting science requirements and mission constraints. Next, the Mission and Sequence Design Team (MSDT) generates the ASPP using the Mission Plan and preliminary ephemeris information from the Navigation Team (NAV) and distributes them to all other uplink teams for review. The Mission Control Team (MCT) then negotiates and schedules DSN station coverage based on activities in the ASPP, while all teams review the ASPP for conflicts and missing requirements. Their findings and requests for change are documented and transmitted to the MSDT.

After the delivery and review of an ASPP, a sequence is built, incorporating any engineering requirements based on prior telemetry analysis, updated DSN support, and any new or modified spacecraft or ground events. A sequence package is produced and once again reviewed by the MOS teams, where requests for change to it are written as necessary and submitted to MSDT. MSDT identifies conflicts among the change requests and distributes these conflict statements to the teams for resolution, and the MCT provides the MSDT with updated DSN station allocations. At a preliminary sequence review, the MSDT presents all change requests, conflict statements, and proposed resolutions to the Mission Director, who approves or disapproves each change. He places any necessary liens on the sequences when violations of constraints are uncovered or when unresolved questions arise. Each preliminary sequence review results in an approved preliminary mapping sequence package which is required to satisfy all margin allocations and constraint checks.

The preliminary mapping sequence package, including time-ordered listings of events, is distributed to all uplink teams. The Spacecraft Team (SCT) receives this sequence file digitally so that it can be constraint checked by computer and expanded to include spacecraft parameters into a command sequence called the preliminary spacecraft events file (SEF). The process of parameter generation and input is a major consumer of time for the Spacecraft Team, carefully performed to ensure that the spacecraft will respond safely and correctly to the planned sequence. This point in the uplink process is the first time that a sequence of spacecraft and ground activities is converted into a sequence of specific spacecraft commands and ground events. This file is meticulously scrutinized by members of all subsystems of the Spacecraft Team.

The MCT uses this sequence to schedule final DSN coverage and to generate a file of antenna-related parameters called a 'keywords file' for the DSN and preliminary versions of two levels of schedules. The latter are used by operations personnel to plan daily activities and by the Operations Planning and Control Team for event monitoring. The highest level schedule is called the Spaceflight Operations Schedule (SFOS). A more detailed schedule, called the Integrated Sequence of Events (ISOE), lists second-by-second events in time order, whether they are local to the operations centre, at the receiving antennas, or on the spacecraft. Sample pages from an ISOE are shown as Figure 7.10 covering the shaded area of the SFOS shown in Figure 7.9. These will be described in detail in section 7.6.

These products (SEF, ISOE, and SFOS) are distributed for review, and any revealed conflicts or new change requests are submitted to MSDT once again. A final sequence review is held next, where all old and new change requests, conflict statements, and resolutions are considered. A final mapping sequence is formed from the approved conflict resolutions, preliminary mapping sequence, final DSN station allocations, and an updated ephemeris prediction (based on more recent tracking information) from the Navigation Team. The SCT once again constraint checks and expands the final sequence into the quasi-final command sequence file. This quasi-final expanded command sequence is reviewed by the SCT subsystems analysts and used by the MCT to generate a final ISOE, SFOS and keyword file. The approved and finalized set of activities produced are then input into the next task, sequence generation.

While the command sequence is being prepared, the Radar System Engineering Team (RSET) receives ephemeris information from the Navigation Team and uses it, together with timing and other orbit-peculiar information, to generate two files: (1) a file describing the roll manoeuvre which the spacecraft must execute during the mapping pass and; (2) a file of radar control parameters. These two files are married with the final command sequence file during sequence generation. The first file defines a manoeuvre expressed as a set of polynomials with the spacecraft clock time as the only independent variable. The polynomials determine the attitude profile to be followed by the spacecraft as mapping data are acquired. The coefficients for these polynomials are packaged into the file.

The second file built by RSET is a file of parameters used by the radar sensor and altimeter to set up the datataking and is prepared for every mapping upload. Due to the rapidly changing altitude and the sensitivity of the radar to exact knowledge of spacecraft position, these parameters must be generated using a recent prediction of the spacecraft ephemeris. Therefore, the timing of receipt of tracking data for processing, generation of ephemeris predictions, and calculation of radar parameters is tightly scheduled and must be executed within that schedule lest the predictions 'age' and become insufficiently precise for the intended datataking sequence. To allow sufficient time for that schedule, the sequence development task proceeds in parallel with it and a file of radar control parameters is delivered directly to the sequence generation task.

Sequence generation, the second step in the uplink process, incorporates the latest spacecraft positional information from navigation tracking by adjusting the timing of the command sequence file to agree. It also verifies the usage and management of on-board memory, ensuring that the sequence of commands will be loaded into the proper locations in memory at the right times. The commands are then translated into binary data format and loaded into a command file. The two files from RSET, discussed above, are received and reformatted to produce command files to be uploaded along with the sequence-related commands.

For Magellan, the sequence generation step also includes the step identified in Chapter 4 as command validation. The Magellan program has a combined hardware and software simulator called the systems verification laboratory (SVL). At the heart of the SVL are breadboard versions of flight computers for command and telemetry handling and for attitude control. These breadboards function identically to their flight equivalents, and can be used for verification of flight software changes. With the rest of the spacecraft simulated with either firmware or software in auxiliary computers, the SVL is a tool for validation of command sequences. During the uplink process, the sequences that require testing in the simulation laboratory are tested at the preliminary SEF level, since there is insufficient time between the production of the final SEF and the time of upload to insert a real-time test.

On the day of the upload, after receiving the final ephemeris data, final ISOE, radar control parameters, mapping attitude polynomial coefficients, and other miscellaneous files, the SCT does its final expansion of the sequence and creates the final command sequence file. A final constraint check is performed, and the commands are translated into three binary files containing the sequence commands, the radar control parameters, and the mapping polynomial coefficients. These files are approved by the Mission Director at a series of upload approval meetings, and transferred by the Operations Planning and Control Team to the multimission command subsystem.

Command processing is the final task in the uplink process. During command processing, the MCT transmits the command files from the multimission command system at JPL through the proper DSN antenna for radiation to the Magellan spacecraft where it is stored for later execution. This action ends the standard sequence uplink process. The command file is typically timed to begin executing about three orbits from the end of the nominal transmission, providing some time margin for the upload if problems with the ground uplink system occur. The ISOE and SFOS are used by the MOS teams to schedule their work and to monitor the spacecraft activities invoked by the command file.

The uplink process is constrained by limited resources on both the spacecraft and the ground. The Operations Planning and Analysis Office and the Radar Office are jointly responsible for limiting the number of sequence uploads being planned and the complexity and density of spacecraft and ground activities in each sequence to that absolutely necessary, thus remaining within the resources (personnel, computer and on-board memory) available.

In addition to the standard sequence process, Magellan has a non-standard command route, originally developed to provide a method of commanding the spacecraft outside of the standard route and more quickly than that route would allow. It generally consists of command loads other than sequences which are executed immediately, but it may include short sequences to be executed at a later time. The non-standard command route was not envisioned for frequent usage based on the repetitive mapping philosophy set forth in the Magellan science objectives, but various realities of operations have required it. As a result, an engineering command review board and several verification steps were instituted to ensure that this route was not abused and that commands approved via this route were thoroughly checked before transmission. Examples of command files that were developed through non-standard commanding are listed below. See Section 7.7 for a description of the anomalies mentioned here.

(1) Turn on the downlink transmitter's travelling wave tube amplifier after a spurious shut-off.
(2) Turn off an erratic gyroscope.
(3) Reset a counter after failed star scans.
(4) Command a contingency star calibration.
(5) Recover from fault protection actions initiated by the telemetry losses.

In general these command files have all been for engineering purposes, have been short, have executed immediately or soon after, and were conceived too late to input via the standard commanding process.

A subset of the nonstandard commands have accompanied standard stored sequence loads. In general these have consisted either of parameter value updates, such as for guide star changes, or commands accessing write-protected memory that can not be implemented via stored sequence but which need to be transmitted at a certain time during sequence execution. Examples of sequence-associated nonstandard commands are:

(1) Changes to parameters used in the filter which rejects noise in the star-scanning process.
(2) Changes to magnitudes, locations and window times for new star pairs to be used for star scans.

(3) Commands to initiate small sequences.
(4) Memory load commands for small sequences.
(5) Fault protection reconfiguration commands to modify the on-board response to certain anomalies.

The standard command route is not very flexible because of Magellan's intent to perform only regular, repetitive mapping. First, the capabilities of this route are, in general, limited to the use of predefined blocks of spacecraft commands, each of which has been thoroughly checked by analysis, simulation and execution on the breadboard spacecraft in the SVL. Commands issued outside of blocks must be either planned well in advance or issued through the non-standard route. Usually the specific commands and parameters are not known far enough in advance to insert the non-block commands into the sequence. Second, requests for changes to standard commanding sent to MSDT in reaction to ASPP, preliminary or final reviews must satisfy the following rules.

(1) The changes may not require the tape recorder data return strategy to be changed.
(2) The changes may not invalidate the *memory management strategy*. This term refers to the organization of the on-board memory where the sequence is stored and the time order of relocating the new sequence in memory to prepare it for execution after the old sequence has terminated.
(3) Changes to a sequence may not affect preceding or succeeding activities.
(4) The number of changes approved must be based on minimizing complexity, manpower, and sequence planning time constraints.

With these restrictions, many necessary commanding activities are moved to the non-standard process. However, in either the standard or non-standard case, the result is performance of the spacecraft and radar instrument of the intended actions resulting in downlinked data.

7.6 *The downlink process*

The major functions and interfaces of the MOS teams in the downlink process are shown in Figure 7.7. The process begins with the DSN Ops Team, which operates the DSN in order to acquire telemetry from the spacecraft and transport these and other data to JPL. Four distinct data types are collected: (1) engineering telemetry, which is intended for monitoring and recording the status of the spacecraft subsystems; (2) user data, which for Magellan means the science data, returned from the SAR, altimeter and radiometer; (3) tracking data, which are measurements of the frequency characteristics of the received carrier signal used to determine position of the spacecraft; and (4) data generated at the active DSN station or stations

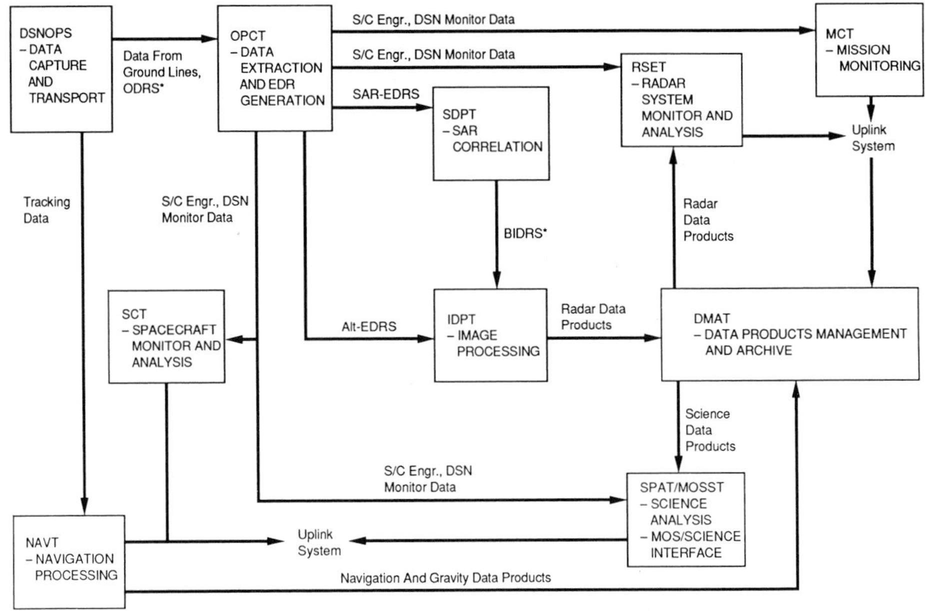

Figure 7.7 Downlink process functions and interfaces.

detailing the status of the equipment used at those stations to acquire and transfer both engineering and science data. This last data type is usually called 'DSN monitor data'.

Engineering data are the only spacecraft data that are monitored in real time. They are downlinked whenever the HGA is pointed to Earth, which during mapping is for two approximately one-hour periods out of each orbit. Figure 7.8 shows the process associated with the monitoring of both engineering and DSN monitor data. The DSN Ops Team personnel at the active DSN station must re-acquire the telecommunications signal and lock up on the telemetry twice each orbit when the high-gain antenna returns to Earth point. Once the telemetry is acquired, it is routed in real time to JPL. Every 5 s, a block of DSN monitor data is injected into this data stream.

The Operations Planning and Control Team (OPCT) operates the computers that receive and process the real-time data. The software executing in these computers creates channels of spacecraft engineering and DSN monitor data to be sent to real-time displays. This processing is based on telemetry processing parameters, such as decommutation maps, decalibration tables and alarm limits, that are generated by the Spacecraft Team. The real-time data are also stored in a computerized database for later trending analysis and for archiving.

Real-time monitoring of telemetry processing is performed by the OPCT as a part of their overall responsibility of monitoring all real-time ground

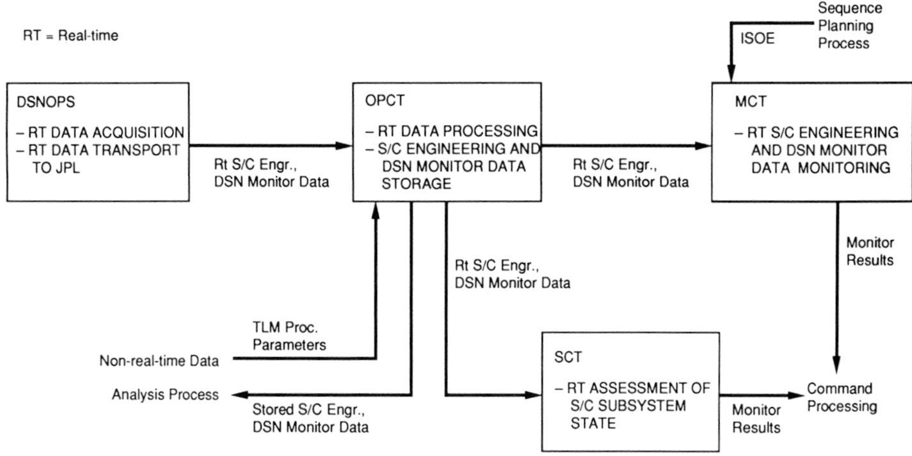

Figure 7.8 Real-time data monitoring process.

processing events. This monitoring is for the purpose of maintaining the ground process and not for identifying the anomalous behaviour of the spacecraft. Their job is aided by the ISOE, which is generated by the uplink process. Real-time channelized data, including alarms when they occur, are displayed on console monitors used by the MCT to monitor spacecraft and ground data system (GDS) events listed in the ISOE; by the SCT, to do a real-time assessment of the spacecraft's subsystem's health; by the RSET, to establish the state of health of the radar sensor; and by other teams to monitor progress of the mission.

Event monitoring is a bridge between uplink and downlink processes in that it compares predicted events to actual occurrences. It is driven by the two schedules generated by the uplink process, the SFOS and the ISOE. The SFOS, shown in Figure 7.9, will be described first. Note from the figure that this schedule covers 24 hr on each page in a graphical presentation. Different classes of events are spread down a page and time increases from left to right. Two time scales are shown across the top, Pacific local time (either Standard or Daylight as appropriate) and Universal Time, Coordinated (UTC), which has replaced the earlier designation of Greenwich Mean Time (GMT). The creation date and time, at upper right, is important to distinguish revisions which are sometimes required due to late updates. The four digit numbers throughout the body of the figure are the times in UTC of each event to the nearest minute, expressed as HHMM, where HH is the hour and MM is the minute.

Progressing down the left-hand side of the figure, round-trip and one-way light time (RTLT and OWLT) are shown, followed by Magellan allocations of the various antennas at the DSN complexes. Each antenna is referred to as

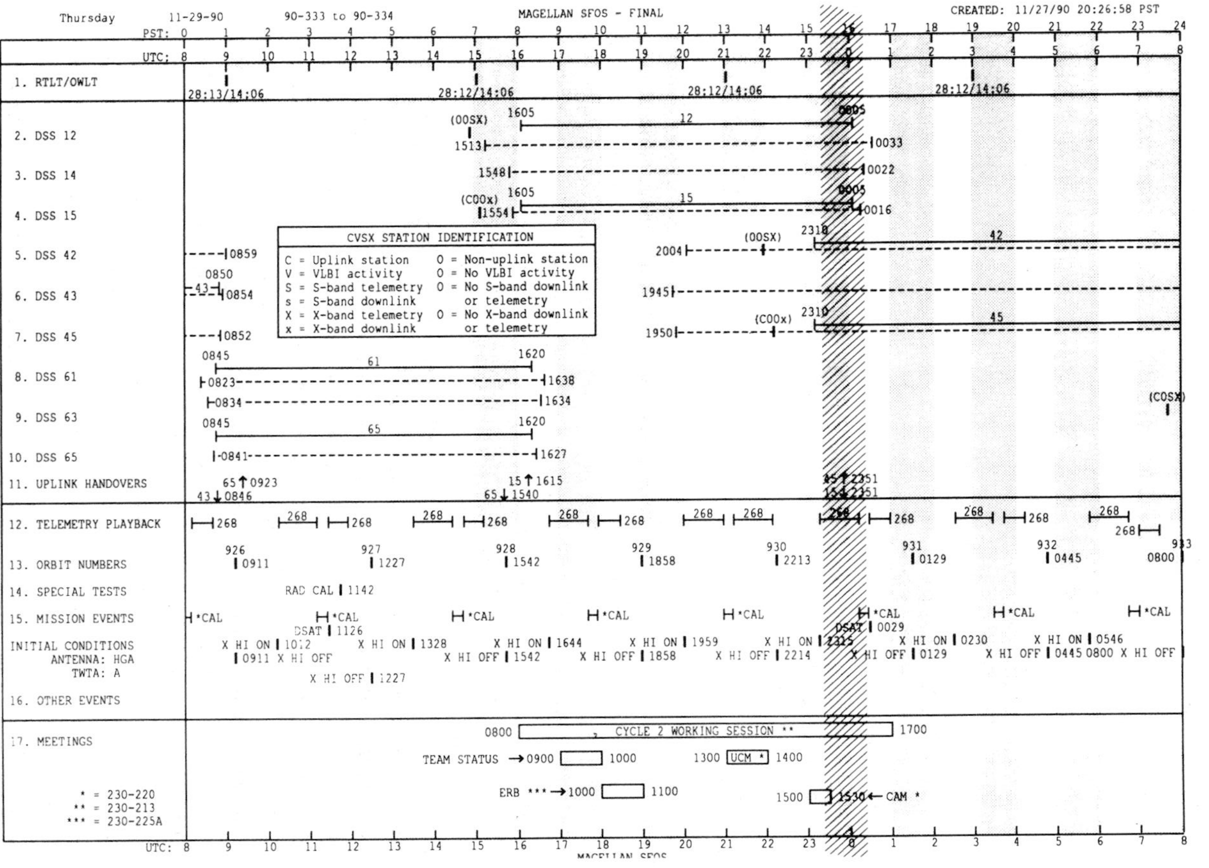

Figure 7.9 Spaceflight Operations Schedule (SFOS) for November 29, 1990. An Integrated Sequence of Events corresponding to the shaded area is shown in Figure 7.10.

a Deep Space Station (DSS) and has a separate number. For example, the large 70-m DSSs (shown in Figure 1.10) are DSS 14, DSS 43, and DSS 63 in the USA, Australia and Spain complexes, respectively. Associated with each antenna is a dashed line for the rise and set times of Magellan as seen from each station and a solid bar for those time periods for which the DSS antenna is allocated to a Magellan contact. Line 11 shows which antenna is available for uplink, and the time at which the uplink responsibility is handed from one DSS to another. Lines 12 through 16 represent various spacecraft activities and configurations, where 268 refers to the downlink telemetry playback rate of 268.8 kbits/s, DSAT stands for momentum wheel desaturation, and *CAL refers to a star calibration for attitude knowledge maintenance. Line 17 identifies the day's meetings which operations personnel must attend. In this example, a session to plan Magellan's second cycle occupies the entire local work day, and team status meetings are held in parallel from 9 am to 10 am local time. Three uplink review meetings called Engineering Review Board (ERB), Uplink Coordination Meeting (UCM), and Command Approval Meeting (CAM) are also listed.

The shaded area in Figure 7.9 is expanded into the ISOE pages illustrated in Figure 7.10. In the ISOE both ground event time (second column) and the associated spacecraft event time (second to last column), different by the OWLT, are given, together with the internal spacecraft clock counts (last column) whenever a line reflects a spacecraft activity. Each event (E) or time (T) is described with a cryptic set of words and algorithms in the fourth column. Many of the events described in this example relate to DSN activities, such as acquisition and loss of the telecommunications signal, vehicle rise and set, and start and ending of tape recording the original data record. Others are spacecraft activities such as turns to and away from mapping, performance of a star calibration attitude update manoeuvre, flight tape recorder position adjustments (6DMSC commands) and changes to on-board software fault protection states. Lines labelled EPS give predicted status of the on-board electric power subsystem. When the event is a spacecraft command execution, the command mnemonic is listed in the fifth column, and when a DSS antenna is involved its number is shown in the sixth.

A separate downlink effort by the SCT also analyses the real-time engineering telemetry, accessing it in non-real time from the database in which it is recorded. Current spacecraft subsystem performance is appended to past performance to establish trends and then extrapolated to predict future performance of subsystems. The values of on-board parameters are carefully tracked, especially those determining fault protection responses and those controlling the performance of the attitude control process in flight software. Current status and predicted trends are used to generate information used in the advanced sequence planning and sequence planning tasks. This analysis may result in updates to on-board parameters or to telemetry processing parameters (such as decommutation or decalibration tables) used by the real-time data monitoring activity. Analysis of the thruster firings from manoeuvres or momen-

ITEM NO.	GND TIME DOY/HH:MM:SS	T E	EVENT - DESCRIPTION	COMMAND	DSN	SCE TIME DOY/HH:MM:SS	S/C CLOCK
02375	333/23:16:27	E	B,P268,1; << DMS CONTROL >>	6DMSC		333/23:02:21	0830985:62:0
02376	333/23:16:27	E	BEGIN RECEIVING HIGH RATE PLAYBACK DATA, NOW, BIT-RATE = 268.8		12	333/23:02:21	0830985:62:0
02377	333/23:16:27	E	BEGIN RECEIVING HIGH RATE PLAYBACK DATA, NOW, BIT-RATE = 268.8		42	333/23:02:21	0830985:62:0
02378	333/23:16:27	E	EPS MB CZ=0.2 CC=0.69 CP=6.04 PRU CZ=0.0 CC=0.0 CP=3.24 INV CZ=0.0 CC=0.0 CP=3.65			333/23:02:21	0830985:62:0
02379	333/23:17:16	E	SDB1A,04F8,186,A; << HIGH RATE MEMORY READOUT >>	6MROH		333/23:03:10	0830986:44:0
02380	333/23:20:16	E	SDB1A,04F8,186,A; << HIGH RATE MEMORY READOUT >>	6MROH		333/23:06:10	0830989:41:0
02381	333/23:21:12	E	MISSION PHASE ORB OPS S/W MODE*ATT REF HOLD	AACS		333/23:07:06	0830990:34:0
02382	333/23:23:16	E	SDB1A,04F8,186,A; << HIGH RATE MEMORY READOUT >>	6MROH		333/23:09:10	0830992:38:0
02383	333/23:44:20	E	A,P268,1; << DMS CONTROL >>	6DMSC		333/23:30:14	0831013:23:0
02384	333/23:44:20	E	EPS MB CZ=0.2 CC=0.69 CP=6.04 PRU CZ=0.0 CC=0.0 CP=3.74 INV CZ=0.0 CC=0.0 CP=3.65			333/23:30:14	0831013:23:0
02385	333/23:44:50	E	DMA,E1,X2S,STOA,M1,XTOA,A1R,T1,DR62, H; << TCS CODED COMMAND >>	2CC		333/23:30:44	0831013:68:0
02386	333/23:44:52	E	B,RDY,0; << DMS CONTROL >>	6DMSC		333/23:30:46	0831013:71:0
02387	333/23:44:52	E	EPS MB CZ=0.2 CC=0.69 CP=6.04 PRU CZ=0.0 CC=0.0 CP=3.24 INV CZ=0.0 CC=0.0 CP=3.65			333/23:30:46	0831013:71:0
02388	333/23:44:54	T	TXR OFF DSS 15		15		
02389	333/23:45:15	E	A; << .9 CAT. BED PRI HTRS-ON >>	4CB1		333/23:31:09	0831014:14:0
02390	333/23:45:15	E	EPS MB CZ=0.98 CC=0.69 CP=6.04 PRU CZ=0.0 CC=0.0 CP=3.24 INV CZ=0.0 CC=0.0 CP=3.65			333/23:31:09	0831014:14:0
02391	333/23:45:17	E	A; << .9 CAT. BED SEC HTRS-ON >>	4CB2		333/23:31:11	0831014:17:0

02392	333/23:45:17	E	EPS MB CZ=1.76 CC=0.69 CP=6.04 PRU CZ=0.0 CC=0.0 CP=3.24 INV CZ=0.0 CC=0.0 CP=3.65			333/23:31:11	0831014:17:0
02393	333/23:58:40	T	ACQ U/L - DSS 45 X BAND		45	334/00:12:46	
02394	334/00:05:00	E	END RECORDING, ARA		15	333/23:50:54	
02395	334/00:05:00	E	END OF TRACK DSS 15		15	333/23:50:54	
02396	334/00:05:00	E	END RECEIPT OF HIGH RATE PLAYBACK		12	333/23:50:54	
02397	334/00:05:00	E	END RECORDING, DDR		12	333/23:50:54	
02398	334/00:05:00	E	END RECORDING, ARA		12	333/23:50:54	
02399	334/00:05:00	E	END OF TRACK DSS 12		12	333/23:50:54	
02400	334/00:08:07	E	D/L RF WARNING, AT 00:13:07 S/C EARTH POINT CHANGE TO OFF				
02401	334/00:13:05	E	A,R7,3; << DMS CONTROL >>	6DMSC		333/23:58:59	0831041:63:0
02402	334/00:13:05	E	END RECEIPT OF HIGH RATE PLAYBACK	6DMSC	42	333/23:58:59	0831041:63:0
02403	334/00:13:05	E	END RECORDING, DDR	6DMSC	42	333/23:58:59	0831041:63:0
02404	334/00:13:05	E	EPS MB CZ=1.76 CC=0.69 CP=6.04 PRU CZ=0.0 CC=0.0 CP=3.53 INV CZ=0.0 CC=0.0 CP=3.65			333/23:58:59	0831041:63:0
02405	334/00:13:07	E	DMB,E1,X2S,STOA,M1,XTOA,A1R,T1,DR62, H; << DMS CONTR ,0.0 ,0.0 ,3.65 ; CMD,7SMAN,280D0459A4A,PRI, 90-333/23:59:00.533,ST2ER,2;	2CC		333/23:59:01	0831041:65:0
02406	334/00:13:07	E	ST2ER,2; << START STAR UPDATE MANEUVER >>	7SMAN		333/23:59:01	0831041:65:0
02407	334/00:13:07	E	MISSION PHASE ORB OPS S/W MODE*SLEW TO ATT	AACS		333/23:59:01	0831041:65:0
02408	334/00:13:07	E	S/C OFF EARTH POINT			333/23:59:01	
02409	334/00:13:07	E	LOS, S		42	333/23:59:01	
02410	334/00:13:07	E	LOS, X		42	333/23:59:01	
02411	334/00:13:07	E	LOS, X		45	333/23:59:01	

Figure 7.10 *Integrated Sequence of Event (ISOE) for the time period shaded in the schedule shown in Figure 7.9.*

tum wheel desaturations which impart a force on the spacecraft results in a file which is used by the tracking and navigation function, to be described below.

Science data are also acquired by the DSN Ops Team as it is played back from the spacecraft. Each one-hour playback period is simultaneously recorded on two original data records (ODR). One copy of each ODR is retained at the DSN. The other copy is shipped to JPL from each overseas DSS weekly (daily for the first 30 days of the mapping phase) and brought in by courier from the California DSN station daily. Overseas shipments take about 2 weeks to arrive. Once the ODRs are received they are checked by DSN personnel and delivered to the project. DSN Ops personnel at the overseas stations also transmit over data lines, three designated orbits' data for use in producing expedited radar data products (see below).

ODR tapes and other ancillary files are used to generate experiment data records (EDRs). Three distinct types of EDRs are made. They are:

(1) The SAR EDR, which contains not only SAR data from the radar sensor but also all ancillary data necessary to turn SAR data into images, including navigation data, sensor engineering data, and information from the radar control parameter generation task.
(2) The ALT EDR, which similarly contains all data necessary to process the altimetry and radiometry data.
(3) The Archive EDR (AEDR), which contains all spacecraft and radar engineering data for placement into the project archive.

The SAR Data Processing Team (SDPT) accepts the SAR EDR and transforms the echoes received from the planet into images. This process is extremely computer intensive compared to other missions and requires special-purpose hardware in order to achieve processing of a day's worth of Magellan data each day (Curlander and McDonough, 1991). From the SAR EDR, the SDPT creates a single strip of image data representing a venusian surface area 20 km wide and about 15×10^3 km long, called the full-resolution basic image data record (F-BIDR). On the F-BIDR, each 8-bit byte contains one pixel (see Chapter 4) which represents a 75-m patch of surface area. Pixels on the F-BIDR are located on the planet in latitude and longitude, calculated from navigation data contained on the SAR EDR. Each F-BIDR is displayed on a long strip of thermal print paper which is delivered to the Mission Operations Science Support Team (MOSST), who checks the strip for image quality before allowing the mosaicking process to begin.

The F-BIDRs are delivered to the Image Data Processing Team (IDPT), which averages arrays of 3×3 pixels into single pixels on the Compressed Basic Image Data Record (C-BIDR), whose pixel widths represent 225 m on the surface of the planet. IDPT also uses location information on the F-BIDR to lay each pixel into image mosaics at various resolutions, creating Full-resolution Mosaicked Image Data Records (F-MIDR) with 75-m pixels, compressed-once mosaicked image data records (C1-MIDR) with 225-m

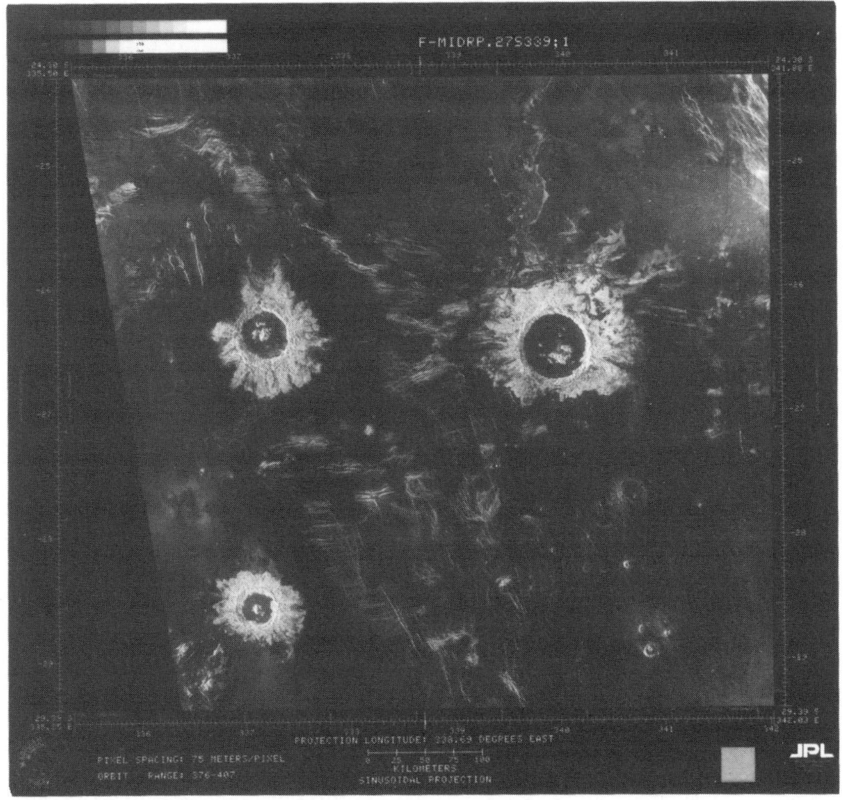

Figure 7.11 Magellan F-MIDR 27S339. MIDRs are assembled from a series of F-BIDR strips (P37375).

pixels, C2-MIDRs with 675-m pixels, and C3-MIDRs with 2-km pixels. These mosaics allow scientists to have both high-resolution and more synoptic views (Figure 7.11). Each MIDR frame is approximately 8000 pixels on each side, covering varying amounts of the surface according to the resolution of each pixel. Each C3-MIDR, for example, covers about one-sixth of the planet's surface. The relationship of these products to each other is shown schematically in Figure 7.12.

IDPT similarly creates altimetry and radiometry data products called the Basic Altimetry Data Record (BADR), Basic Radiometry Data Record (BRDR), and their global counterparts, the GADR and GRDR. IDPT also generates more specialized data products on request from science team members.

All science data products are delivered to MOSST, who quality checks products before they are provided to the science teams for analysis. Finally,

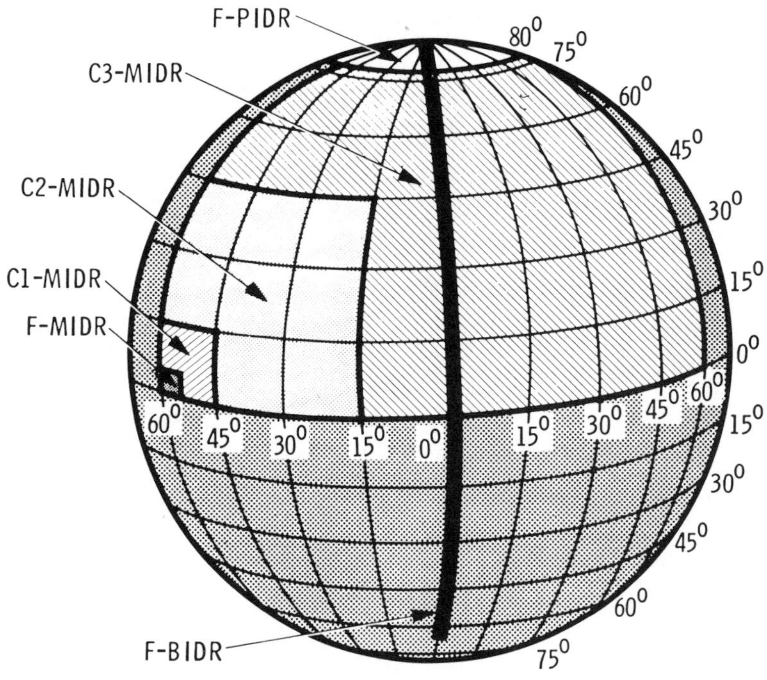

Figure 7.12 Magellan SAR image products, shown in relation to each other.

the Science Planning and Analysis Team (SPAT), consisting of NASA-chosen scientists, analyses data products.

The DSN Ops Team also operates the DSN to acquire the tracking and navigation data used to determine Magellan's trajectory and to refine the gravity field of the planet. Data concerning signal Doppler shifts are collected and transmitted to JPL in real time. Calibration information associated with the Doppler observations is also collected by each station and transmitted along with the Doppler data. At JPL, DSN Ops personnel provide corrections to time and terrestrial polar motion. These data are delivered to the Navigation Team (NAV). NAV uses these files, together with data concerning small platform-created forces (determined from thruster firing times) from the SCT, to determine the trajectory of the spacecraft.

After an initial set-up period in the morning, the daily navigation solution is made based on an iterative set of orbit determination and trajectory programme runs. Once the navigation solution has been completed, ephemeris information from it is used to process images, as described below. The navigation solution is also used to predict the position of the spacecraft at future times, for use in the uplink process and in temporary radar products, described below. Navigation data is also sent to SPAT's Gravity Investigation Group, who use the ephemeris and the thruster data from the

SCT to discover small deviations of the spacecraft from its intended track due to mass concentrations within the planet. These data can, in turn, be used to produce models of Venus' interior by comparing the locations of mass concentrations to topographic variations.

In parallel with the above product flow, smaller amounts of science data are transferred to JPL from DSN stations by alternate paths. These data are processed into EDRs and F-BIDRs as well, but their delivery is accelerated so that the RSET may validate the radar command process and, to some extent, the radar sensor health with them. Two such product types exist. First, products termed 'temporary' are created from some orbits or portions of orbits downlinked to the California DSN complex. These parts of the data stream are extracted from the same ODRs that are normally brought by courier to JPL from the complex, but the OPCT creates special EDRs called SAR TEDRs and ALT TEDRs from them using predicted ephemeris data generated by NAV. TEDRs are generally available sooner than EDRs because they do not have to wait for the actual NAV ephemeris data. TEDRs are processed into F-TBIDRs, which are checked to ensure that sufficient coverage overlap exists between them. Miscommanding of the radar could possibly cause lack of overlap, which would create a discontinuity in the resulting mosaic, and the temporary products allow such miscommanding to be identified before too many discontinuities occur.

The second accelerated product type is termed 'expedited'. Expedited products are made from Magellan orbits which are downlinked to overseas DSN stations. Data from selected orbits are transmitted at reduced rates to JPL and are processed similarly into Temporary EDR-like products called XEDRs, again for the purpose of detecting anomalies in the science data in advance of the standard science data path. F-XBIDRs, analogous to F-TBIDRs, are made from SAR XEDRs and checked by RSET.

The key to the operation of the downlink process is the Data Management and Archive Team (DMAT), who manages and tracks all data products used by the project. Every data transfer mentioned above is actually handled by DMAT, who expedites those products contained on transportable media (such as computer tapes). DMAT also records the locations of all products and is responsible for cataloguing their status in their internal database. DMAT operates the working library that makes the data products available for local users, and manages the project archival facility that contains archive copies of the data products. Responsibility for delivery of the final project archive to the NASA planetary archive also lies with DMAT, under the direction of the Project Scientist.

7.7 *Anomalies*

At the end of the prime mission, Magellan has encountered three types of major anomalies, two of which involved significant losses of downlink telemetry, and several additional minor anomalies. Fortunately, none of these has seriously impacted the science return of the mission, although both have had long-lasting effects on the operation of the spacecraft. The major anomalies to date, their impacts and their resolution have been as follows:

(1) ANOMALY: An address bit on either one of two chips in the back-up attitude control computer's random-access memory intermittently fails. When this bit fails, 2048 bytes of memory (out of 32768 bytes) cannot be read or written.

IMPACT: No science data have been lost because this problem has been contained within the backup computer. However, currently the mission does not have a functioning back-up attitude control computer in the event the prime computer fails.

RESOLUTION: A temporary modification has been made in the action the spacecraft would take in the event of a failure of the prime attitude control computer. Instead of switching autonomously to the back-up, the new action is to stop mapping, turn off all nonessential power loads, and put the spacecraft into a safe, commandable state where it will await instructions from the ground. Meanwhile, a group of flight computer experts are rewriting the attitude control flight software to function in 4096 bytes less memory. Once this is accomplished and verified on the ground, an upload to the spacecraft will restore full redundancy in flight computers.

(2) ANOMALY: An occasionally recurring corruption of data on the main data bus of the primary attitude control computer causes the computer to spuriously transfer processing to an unintended memory address.

IMPACT: This problem has occurred five times since the spacecraft achieved orbit around Venus. The first two times caused a delay in the start of mapping operations. The last three times caused small outages in the data until contact was re-established. Since the specific address to which processing is transferred is random, the impact can vary from none to the cessation of mapping until the ground can take over and recover.

RESOLUTION: The root cause of this anomaly was eventually discovered to be a flight software timing interaction between a foreground and a background process in the attitude control computer. During a 7-microsecond window in each orbit, a foreground pointer address was subject to being overwritten by the background process, an eventuality that occurred on 5 of the 1789 orbits in Cycle 1. A software patch was designed, tested and planned for upload to the spacecraft in July 1991 to fix the problem.

(3) ANOMALY: The star scanner occasionally detects false stars, probably due both to particles of insulation floating in front of the star

scanner which reflect light to the detector, and to solar protons of 10 to 50 Mev striking the detector and causing a spurious signal trace. The star scanner is used to update the spacecraft's attitude knowledge, and the false signals sometimes lead to incorrect attitude updates, thus slightly mispointing the antenna.

IMPACT: The magnitude of an incorrect attitude update is limited by checks performed in onboard software to a value that will not cause loss of telecommunications. This worked well during cruise, where several bad updates were accepted with no significant impact. However, in December 1990, two spurious updates occurred only a few hours apart. The sum of the two updates exceeded the limits, causing a loss of signal and loss of some mapping data.

RESOLUTION: A number of flight parameters were adjusted to better define the time window of an acceptable star crossing in the detector as well as the range of acceptable star magnitudes. In addition, steps were implemented on the ground to recognize and respond to one incorrect attitude update as rapidly as possible before a second one could worsen the situation.

Notice that all three of the above anomalies have resulted in a greatly increased workload on the flight team. Additional flight software, additional ground procedures and additional command loads to the vehicle are common to all three. This is a characteristic of in-flight anomalies and emphasizes the need for operations teams to anticipate them.

Several less significant anomalies have occurred as well. They are discussed below in the same style as above. As above, all put extra burden on the ground resources.

(4) ANOMALY: The telecommunications subsystem's travelling wave tube amplifier (TWTA) experiences spurious shutoffs (SSO), which cause either an automatic switch to the backup TWTA, an automatic restart of the same TWTA, or in certain circumstances, a shutdown of the X-band telecommunications subsystem.

IMPACT: Several SSOs have occurred, and the impacts have been minimal. When TWTAs are swapped, the downlink circular polarization is reversed in direction. In general, only four minutes of data are lost while the traveling wave tube warms up and the DSN changes polarization. One SSO resulting in a TWTA swap caused 20 minutes of data loss because the DSN did not immediately recognize what had happened and delayed switching polarization.

RESOLUTION: Flight software was modified to ensure either a TWTA restart or a switch to the redundant TWTA so that X-band telemetry will suffer minimal interruption. An addition was made to DSN procedures to try the opposite polarization directions in the event of difficulty in acquiring the signal.

(5) ANOMALY: Noisy and excessive current was drawn by one of the gyroscopes.

IMPACT: There has been none on science data return. Significant activity has been required on the ground to understand any possible effect on the gyro's lifetime.

RESOLUTION: The noisy gyro was turned off and the gyro sets were redefined so that the noisy gyro would only be turned back on as a last resort (gyro sets are doubly redundant).

(6) ANOMALY: Spacecraft component temperatures have been higher than anticipated, due primarily to more rapid degradation of properties of thermal surfaces than expected.

IMPACT: The life of some flight components has possibly been shortened. There has been no impact on prime mission science, but data collection during the second cycle will be significantly impacted due to thermally difficult spacecraft attitude during tape recorder playback back periods.

RESOLUTION: Cycle 2 science data collection strategies have been redesigned to account for much more severe thermal constraints.

(7) ANOMALY: A pattern-sensitive corruption of data is occurring during playback from tape recorder A where a 0001001 pattern is occasionally being changed to 000111, causing a short frame.

IMPACT: Corrupted data are very difficult to process into images. The SAR processor fails to process data surrounding the short frame.

RESOLUTION: Two activities are happening in parallel. Ground software is under development to recognize and correct the data corruption, and tape recorder tests (on both the in-flight recorder and a spare on the ground) are being planned to investigate the cause of the problem and correct it on-board.

7.8 Summary

The Magellan mission has been described in detail in order to show an example of how the general principles outlined in other chapters may be implemented. Magellan, like most remote-sensing missions, has many aspects that make it unique, but its mission operations system is typical in many respects. Magellan was conceived as a low-cost, non-adaptive mission whose objective is to map the surface of Venus using synthetic aperture radar images, microwave altimetry, and passive microwave radiometry. All three of these measurements are taken using a single instrument. In addition, the spacecraft itself is used to map the venusian gravity field through accurate tracking of its orbit.

Table 7.1 Key terms from Chapter 7

cycle	Project Science Group
forward equipment module	radar sensor
high gain antenna	Universal Time, Coordinated
Magellan	Venus Orbiting Imaging Radar

The Magellan uplink and downlink processes have been described together with the teams that operate them. Anomalies that have been experienced to date are listed, together with their impacts and resolutions. Although none of these anomalies has threatened the mission's capacity to accomplish its goals, they have significantly altered the way in which the spacecraft and its instrument are operated by the MOS and its personnel. Table 7.1 lists the key terms introduced in this chapter.

In the next chapter we will see how recent technological innovations will make future MOSs operate more smoothly and at lower cost.

References

Elachi, C., 1988, *Spaceborne Radar Remote Sensing: Applications and Techniques*, New York: Institute of Electrical and Electronics Engineers.

Saunders, R. S., Pettengill, G. H., Arvidson, R. E., Sjogren, W. L., Johnson, W. T. K. and Pieri, L., 1990, The Magellan Venus radar mapping mission, *Journal of Geophysical Research*, **95**, B6, 8339–8355.

Curlander, J. and McDonough, R., 1991, *Synthetic Aperture Radar: Systems and Signal Processing*, New York: John Wiley and Sons.

8

The Future of Mission Operations Systems

To predict the future in any field is difficult. However, recent changes in the way mission operations systems are being designed, with the objective of reducing overall cost, provide strong indications of the path to the future. This chapter discusses five trends that will forge this path: (1) the use of high-speed, high-reliability data communications systems to distribute operations functions to more convenient remote sites; (2) the growing ability for the user to operate a sensor remotely through direct interaction without involvement in cryptic computer commands; (3) the continuing rapid increase in memory and processing speed of flight-qualified computer chips, making in-flight systems more capable; (4) the advent of artificial intelligence in the form of special software systems to augment ground functions; and (5) the expansion in automated, model-dependent science data processing to simplify handling of the vast increase in the quantity of data being returned.

All five of these trends will be significantly aided if the current movement to standardize certain aspects of telemetry and command system design is successful. This chapter will conclude with an examination of this standardization.

8.1 Remote operations: distributed systems

In the past, scientific remote-sensing missions have been operated from a single site, the mission control centre, where the people involved in operating the platform and its payload instruments congregated. This site contained the communications equipment and mainframe computers that performed command and telemetry functions and provided number-crunching capability for spacecraft tracking, subsystems performance analysis and sensor data processing. It was extremely advantageous to have everyone connected with the mission available in person to conduct face-to-face meetings, review mission plans, and discuss interpretation of findings. It was also tremendously expensive.

During the Viking Mars missions of the 1970s, several hundred engineers and scientists converged on JPL at Pasadena, California to participate in mission operations. Many of these were engineers from contracting companies that built components of the spacecraft. They were there to support the operation of their component and needed to stand by in case something went wrong.

Many more were university scientists and professors who were there to support either instrument operation or analysis of experiment results. The majority of these were away from their homes and were reimbursed from project funds for incurred travel and living expenses. In the case of teaching professors, a substitute had to be hired at the home institution to fill in during their absences. All of this travel-related expense consumed large quantities of project funds.

Then in the 1980s, the small, powerful desktop computer with a communications modem made its entrance and the spacecraft operations world has not been the same since. Given sufficiently capable computer equipment, remotely distributed but connected through reliable high-speed data lines to a central source, and consistent, reliable voice communications, there seems to be little reason for ever leaving the convenience of home base. The potential yield in cost savings of remote operations is great. For the engineers specializing in one component, local access to their data and the ability to input into the operations process from home base means not only elimination of additional travel costs, but the possibility of part-time support to the mission only as the level demands, further lowering incurred costs to the project. In like manner, the scientist not only eliminates extended travel costs but may be able to use the resources of the home university, including inexpensive student assistants, to perform data processing and analysis.

Goddard Space Flight Center has been examining spacecraft control centres with remotely distributed functions. The typical GSFC mission is an Earth-orbiting science satellite, built by a prime contractor from industry and containing several independent, but related, experiments designed by university principal investigators. Studies have been done to investigate the potential for leaving the non-institutional support personnel at their home location. One such study (Ledbetter and Taliaferro, 1983) attempted to assess the distributability of each function performed in a command and control centre operating a typical mission. Some of the results are discussed below.

A *distributed operations system* is defined as an operational system where one or more functions are physically removed to a location remote from any other function so that all voice and data interactions must be accomplished indirectly through an electronic medium. The decision of whether or not to distribute a function requires analyses of the types discussed in Chapter 2, specifically a multilevel functional decomposition followed by a detailed analysis of the interfaces between functions. For a function under consideration for distribution, both the quantity of its interfaces and the complexity of each interface must be considered. In most cases, functions such as real-time commanding or spacecraft tracking should remain centralized, since most missions utilize an existing multi-mission command, telemetry collection and tracking network, such as GSFC's TDRSS or JPL's DSN, and the existing interfaces are both numerous and complex.

Studies have shown that functions performing coordination between the other elements usually need to remain centralized. A platform with several different instruments will require coordination of resources used by the experiments in performing their desired measurements. For example, the sum of the power requirements of all experiments at any given time cannot exceed the capability of the spacecraft power subsystem to provide. Two optical instruments firmly mounted on the same scan platform cannot be pointed in different directions at the same time. These types of issues are settled by a central conflict resolution portion of the mission planning functions, as we have discussed in detail in Chapter 4.

Proper distribution of functions in this scenario might be as follows. Remotely-located experimenters electronically submit their observation requests for their instruments according to the approved mission plan and compete for observation times and resources needed from the platform. The central coordination function receives all requests, merges them, identifies conflicts and contacts those experimenters whose plan needs to be modified due to conflicts. The central coordinator identifies options for adjustments to the observation requests that will solve the conflict and then works with each experimenter to select the optimum solution for the project. More than one iteration may be required. Part of the step to identify conflicts usually involves transmitting a file of requested resource usage data to the spacecraft engineers at their remote site for analysis. Results are returned to the central coordination site. Eventually a coordinated schedule of activities is released by the central function.

Based on studies of distributability of operational functions and experience gained from remote operation of the Magellan spacecraft engineering team, the following thirteen suggested implementation approaches are keys to making distributed systems work efficiently. The order does not imply any priority.

(1) Dedicate a central element to the function of coordination of payload experiment operations and production of an integrated schedule. This function becomes a node for all commanding requests, greatly increasing the probability that coordination and conflict resolution will be efficiently performed. The alternative, distribution of this function, creates an unmanageable maze of additional interfaces.

(2) Assign one single element the responsibility of maintaining and modifying a given database. Different elements can be responsible for different databases. In a highly automated, remotely distributed system, database control becomes increasingly important since discussion between users is less likely to occur. Incorrect assumptions about what other users have or have not entered into a database can lead to operational errors. The ability to change or delete information in each database must be restricted to a single controlling source and requests for change must be channelled through that source.

(3) Disseminate information electronically to all operational elements through external read-only database access and transfer. The source function updates the database and time tags the update or adds an indication that a revision has been made. The using function accesses the database on a read-only basis whenever the information is needed and checks the currency of the data. If desired, the user can copy the data to a file or printer for retention, but he cannot modify them.

(4) Automate as much as possible of the constraint checking and conflict identification process. This is especially true of the tasks performed by the central coordination function because of the potential for large numbers of platform or payload activity requests converging at this point for integration.

(5) Perform constraint checking and conflict identification as early as possible in the sequence development process, beginning with checks for logic, content and potential interactions at the investigator's site prior to submittal to central coordination. The earlier that conflicts are discovered, the easier they are to fix. If the experimenter can execute software that will check his observation request file against a current platform and payload schedule database, many conflicts can be resolved at the source before being input to central coordination.

(6) Establish access codes and other security measures for protection of data. If the central function cannot physically see the remote user who is logging on to the system, there is no guarantee that the user is authorized. The end result could be detrimental to the mission. Extensive use of callback systems, where a remote user calls in and identifies himself before the computer hangs up and calls him back at an authorized location, and other standard computer security techniques are highly recommended to prevent intrusion.

(7) Establish communications interface standards well in advance of hardware selection. A set of standards and protocols must be developed (or selected from commercially available sets) for communications with remote experimenters or spacecraft engineers, to guide them in the selection of compatible equipment. This eliminates the difficulties that arise when trying to network together different hardware and software systems, and will increase the efficiency of operations.

(8) Define the content of planning and scheduling products and their operational interactions well ahead of the start of the mission operations phase. This includes descriptive details of the methods remote users will use to modify their experiment plans, submit observation requests and obtain appropriate approvals.

(9) Define, in advance, detailed work-around and recovery procedures for the loss of a remote data link. Operations procedures must carefully

delineate the steps to take if a remote data link is lost. Back-up methods of transferring information between central and remote sites should be designed and tested. Prestored and prevalidated spacecraft or instrument safing commands should be prepared and stored at the central (or upload) site to cover emergency situations where a line could be lost at a critical time.

(10) Design operations software to be modular and flexible, using common higher-order languages and operating systems that have been proven. Achieve as much commonality between users as is possible within the constraints of individual experiment goals.

(11) Design a voice communications system that provides quick, clear query and response capability between remote and central operational positions. Misunderstandings due to poor communications equipment can lead to incorrect commanding.

(12) Plan for and install a quality teleconferencing system. Although meetings are universally disliked, the core of mission operations is usually a series of regular meetings where decisions are made concerning conduct of the mission. Inability to fully participate in these decisions due to poor teleconferencing equipment leads to fragmentation of the flight team. Also give due consideration to the time of day teleconferences are called. A meeting scheduled near the end of the work day for a site in one time zone can be a bit of a hardship for those in a distant zone.

(13) Produce regular status summaries for the rest of the project covering the accomplishments at the remote sites. Sometimes the old adage 'out of sight, out of mind' really applies, and when operations activity becomes tense, the pressure can cause one location to believe the other is not holding up its fair share of the workload, resulting again in flight team fragmentation. Although everyone works their best for the good of the mission, it is always best to communicate what is being done.

The above list is not in priority order, and in fact the last three items relative to communications are perhaps the most important on the list. Good communications are important to efficient operations regardless of the physical location of elements of the flight team, but they are crucial when the participants in different locations cannot hear each other well and cannot see the body language that is so essential to interpretation of expressed sentiments. This point is emphasized by the Magellan experience with the 72-person Spacecraft Team in Denver, Colorado and the rest of the flight team in Pasadena, California. Until an upgraded audio system was installed in conference rooms in both locations, there was a tendency for unhealthy we-versus-they attitudes to develop.

In conclusion, it is probable that distributed control of both platforms and payload experiments will be the way of the future in mission operations, due chiefly to three Cs: cost, convenience and computer capability. Efforts to reduce the cost of performing mission operations will demand it. Operations personnel will enthusiastically support the convenience of staying near their employers, homes and families. And the continued, nearly exponential, increase in the capability of small computers will make it all possible.

8.2 Remote operations: telescience

A concept called telescience originated in the late 1980s within NASA which envisions allowing the user to operate a sensor remotely through direct interaction without involvement in the distracting syntax of coded computer commands. A further advance of the distributed operations system discussed above, telescience has as its ideal a remote-sensing system which is as convenient for the scientist to operate as is a piece of laboratory equipment. An idealistic telescience operations scenario might be as follows. The scientist conceives of an idea for an experiment with an in-flight instrument, walks into his lab, and sets up the experiment himself. Then, depending on the nature of the experiment, either hours, days or weeks later, he sees results. If he wishes to take a picture of a target, he selects a field of view using a joystick, watches a computer-generated version of the target pan by on a screen in his laboratory, zooming in and out as desired, and clicks the shutter himself when he has identified an item of interest. His shutter click is automatically transformed into the commands necessary to acquire the image he desires. He has not been required to participate in, or even know about, the conflict negotiation, resource management, command uplink, telemetry downlink, or data processing. He has no concern or need to know about such details as handovers between tracking stations, on-board memory allocation or other operational concerns because they are made transparent to him. When the data acquisition is completed he is notified, and he is free to roam through the collected dataset, request and receive various products, again without reference to operational considerations.

This concept of telescience is complex and difficult to implement. With a major mission it is probably not able to be realized as described, at least until very high levels of automation and ground system dependability and very complex real-time conflict resolution mechanisms can be developed. Nevertheless, the closer one can approach the above situation the more efficient can become the process of data acquisition. The telescience concept has been broken down into telecommanding, where remotely generated science experiment requests are automatically collected, checked for violation of policy or mission rules, prioritized, conflict resolved, sequenced and readied for uplink; and teleanalysis, where data products are requested remotely and transmitted to analysis centres over high-speed communications lines.

To become a reality, telecommanding requires significant improvements in several functional areas of the mission operations system, some of which were addressed in the preceding section. First, security systems must become so effective that it is prudent to allow the remote collection of user inputs (see items 3 and 6 in Section 8.1). Second, the automation of constraint checking must be able to outdistance the ever-increasing demands for foolproof and error-free commanding (see items 4 and 5 in Section 8.1). Third, all portions of the sequencing and commanding tasks which lend themselves to automation must become automated. Finally, as conflict resolution becomes a highly automated activity it must also become highly efficient — that is, the number of requests denied due to irresolvable conflicts must come very close to the theoretical minimum.

The second element of telescience, teleanalysis, generates additional demands on ground communications capabilities. Today's missions generate vast amounts of science data: unprocessed data from the complete Magellan map of Venus from the prime mission will consist of more than 3 terabytes, and the proposed Earth Observing System will generate several gigabytes each day as currently conceived. An inherent advantage of keeping the mission library and user in the same location is that 'browsing' the data is made easier with the higher data rates possible between archive and local workstation. Using images as an example, an investigator can ask to see a (literally) global view, subsampled to fit the display screen, then narrow the field of view and increase the resolution until the object of interest is visible at the highest resolution available. From a remote location, this function is much more difficult to provide in a reasonable amount of database access time, although the past few years have seen increases in telecommunications rates that may make it feasible in the near future. The Mars Observer and SIR-C missions both plan to implement limited versions of teleanalysis by combining high-speed communications lines with state-of-the-art data compression techniques and processing systems that are optimally distributed between operations centre and users' home institutions.

With minimal additional advances in the state of the art in command integration, constraint checking, and real-time feedback response, and in data handling, compression and processing, the benefits of telescience may be on the near horizon — and the associated benefits may be beyond what we can presently imagine.

8.3 Smart on-board systems

In the 12 years between the 1977 Voyager launch and the 1989 Magellan launch, the state of the art in on-board computer memories for planetary missions took a great leap. Voyager flew with the equivalent of 34 kbytes of 8-bit byte memory in 3 processors (2 of 3 used 6-bit bytes) whereas Magellan has 160 kbytes distributed over four components of the command and data system computer and another 32 kbytes in the attitude and articulation control computer. Since every computer component in both missions was redundant, the total accessible memory increased from 68 kbytes to 384 kbytes. There is every reason to expect that this expansion of flight-qualified computer memory size will continue, and processing speed along with it. What are the implications for future mission operations?

The extra memory on Magellan was used in two ways, both applicable to other spacecraft and missions. One use was to increase the sequencing memory area so that longer command sequences could be loaded on-board less frequently. Had Magellan been required to live within Voyager memory allocations, it would have needed to be commanded much more frequently, perhaps on a daily basis, to ensure collection of data for an adequate radar map of the planet. The resultant increase in mission operations costs may well have been prohibitive.

This use of expanded memory reduces the number of times the ground must transmit commands and therefore reduces both the cost of mission operations and the risk to the vehicle from potential bad commands. The second use of the expanded memory was to develop and load additional fault protection software on-board to assume control of the spacecraft in the event of certain anomalies. There was room in Voyager's memory for a very limited set of fault protection algorithms: those that performed simple component or string switches in the event of a failure, and safed the spacecraft if the problem was more involved. Magellan, on the other hand, had more complicated logic which made use of a myriad of fault protection flags, either maintained by the flight software or updated from the ground, to select a correct response to a variety of different fault conditions. It is worth noting that management of this additional memory and its contents, and prediction of fault protection interactions, adds significant complication to activity on the ground. This added complexity is partially mitigated by the elimination of operations activities that would be necessary if Magellan did not have the ability to autonomously respond to anomalies.

Given a typical spacecraft with multiple vehicle states, hardware component redundancy, several instruments and a mission with multiple phases, the number of potential outcomes from a component failure or other anomaly is very large. Fault protection software can be developed which will take into account these many configurations and protect the vehicle and its components (including experiments) from catastrophic failure. This is especially important for planetary missions, where the round trip light time adds such a transmission delay between on-board problem occurrence and any problem-solving commands that it precludes ground control involvement in rapid response to anomalies. In fact, fault protection flight software could probably be designed that would consume any remaining available memory no matter how much was available. A challenge to future mission designers is to incorporate an optimum level of fault protection into flight memories so as to balance the need for on-board reaction to problems and their probability of occurrence against the code complexity, cost and memory required for implementation.

Another area for potential expansion of on-board capability is in the instruments themselves. Additional memory and speed can allow sensors to select certain observation parameters themselves rather than depend on ground commanding for direction, much like modern automatic cameras select proper exposure time, shutter opening and focus based on sensed information as opposed to older designs where everything had to be manually adjusted. Instruments may also be designed to react to unexpected events or phenomena in a way that would be impossible from the ground. Instruments are also becoming more reliable since mechanically moving parts in sensors, traditionally the most prone to failure in a space environment, are quickly being replaced by their electronically sophisticated equivalents.

The goal of the 'smart sensor' is to provide the most information with the lowest number of both uplink and downlink data bits. Data compression, in both uplink and downlink telemetry (as discussed in Chapter 5), provides the beginnings toward such a goal. With the advent of low-level instructions coded directly into an instrument's memory, automated data acquisition sequences can be contained within the instrument itself. These can be invoked to automatically begin data acquisition if motion is detected, when sufficient light is available, or when other specific conditions occur. The spacecraft must have the ability to store data thus acquired on board, since the ground station may not be in view when the proper conditions happen.

In the future, instruments will contain many different hardware-integrated data compression schemes which can either be chosen by ground command or selected autonomously by the sensor, based on the content of the raw data (Huck *et al.*, 1990). Acousto-optic tunable filters can tailor spectral response of solid-state imaging sensors, for example, and could take an image in which the brightness of image pixels correspond not to amounts of red, green or blue but rather to the amount of a specified mineral contained in the spectral return of each element of the imaged scene. New methods of data compression, already proposed by some NASA workshops, would reduce downlink usage. For example, an algorithm may compare data from an imaging spectrometer to an on-board dictionary of target spectra. If a pixel's spectrum has been entered in the dictionary, only a code pointing to an identical dictionary on the ground is transmitted. If the spectrum is new, the entire spectrum is transmitted and a new entry is placed in both dictionaries for future use.

As faster and more capable space-qualified computer memory and processing units become more available and less expensive, the load on the uplink and downlink systems can be eased by moving more of the complexity aboard both platforms and payloads. However, this is true only if the MOS designers and spacecraft designers plan in advance to take advantage of it. Without advanced planning, the burden on the MOS may actually increase due to a more complicated in-flight system to manage.

8.4 Artificial intelligence applications

Artificial Intelligence (AI) is a field of computer science concerned with the concepts and methods of programming computers to make decisions based on information, logic and deductions that are generally recognized as properties of human intelligence. Branches within this field include robotics, automatic programming, expert (or knowledge-based) systems, natural language processing, and electromechanical visual recognition and/or perception. Although there may be future uses for AI in on-board systems, the more immediate application is to increase the level of automation of ground software systems, thereby decreasing the dependency on humans and saving cost of operations. Of the AI branches listed above, the expert system has the most applicability to ground systems that support remote-sensing missions (Hyman, 1984).

An *expert system* is a computer program that performs a specific task by operating on a large knowledge base using symbolic inference and logical rules to arrive at a statistically probable answer to a query or problem. The knowledge base contains the collected knowledge from human experts in the field, transformed into machine readable form such as data statements and relationship rules. In current AI efforts, expert systems result from the application of several core research topics including techniques for modelling and representing knowledge; the automation of common-sense reasoning and logical deduction; problem solving; and techniques for *heuristic* search. The term *heuristics* refers to rules learned through discovery, and may be in the form of if-then-else rules with associated confidence factors.

An early application of expert systems was a program called DENDRAL, built to interpret data from mass spectrography. DENDRAL was described as an intelligent assistant because it generally produced adequate hypotheses to explain input data. The logic to accomplish a working hypothesis was not programmed into DENDRAL, but instead was derived by the software from its database containing hard knowledge on the properties of chemicals, rules about the outcome of various analyses, and other rules on how to apply the rules (Hyman, 1984). Another area where expert systems have been gaining widespread respectability is in medical diagnosis. Programs such as CASNET and MYCIN contain large databases of medical illness and disease symptoms, properties of medicines, and historical usage results. They are invaluable in assisting doctors in diagnoses, since no single person can retain such large information bases in memory.

The applicability to spacecraft mission operations systems for remote-sensing missions appears to be highest in areas of component design, planning, scheduling, component status determination, problem diagnosis, data analysis and interpretation, mathematical analysis, and ground system configuration maintenance. In the early 1980s, both JPL and the GSFC began development of prototype expert systems in several of these areas, designed specifically for spacecraft operations applications. While those prototypes developed at JPL have not been used within formal operational software builds on actual missions in flight, some of them have been used in parallel with normal flight operations activities on the Voyager project to evaluate how well they performed.

The first area of expert system development at JPL was for a prototype spacecraft activity scheduler. The initial effort was a program called DEVISER. It was capable of using detailed experiment goals and a statement of the spacecraft initial state as inputs to perform scheduling and conflict resolution and to produce a time-ordered sequence of events as an output. This was accomplished through analysis of the end goals achieved with respect to various means employed, to generate a structure representing a least-commitment solution to the sequence. Its major feature, not utilized by previous attempts at automatic scheduling, was the modelling of time in the goal analysis scheme (Rokey and Grenander, 1990). The experiments using DEVISER running in parallel with the standard manual scheduling process on Voyager indicated that it might be able to reduce the workload of scheduling personnel by as much as 50 per cent (Vere, 1981; Wallis, 1982).

In 1986, the experience and knowledge gained from DEVISER was incorporated into PLAN-IT (Plan Integrated Timelines), a tool used by Marshall Space Flight Center in early scheduling efforts for Spacelab Shuttle flights. The JPL Sequence Automation Research Group also applied PLAN-IT to several other scheduling tasks for such diverse activities as antenna scheduling for the DSN, scheduling space station power system usage, and in a joint demonstration with GSFC in telescience applications. Then, in 1988, the lessons learned from these activities were folded into the second generation PLAN-IT-2. This new version is a planning and scheduling tool able to sequence at multiple levels, from experiment goals, through the activities required to obtain these goals, all the way to the step level where the fine details of the experiments are performed. Its operation is controlled by five different types of independent processes that can execute in parallel. PLAN-IT-2 incrementally develops the sequence through user interaction. It incorporates the ability to dynamically add, during the scheduling process, new goals or new semantic-formatted actions to perform. (Eggemeyer and Cruz, 1990)

The second expert system application for spacecraft operations at JPL is for telemetry monitoring and analysis. The initial program was called FAITH for Forming And Intelligently Testing Hypotheses. The purpose of this experimental system was to diagnose failures in both spacecraft and ground control systems. FAITH employed both empirical and system simulation techniques working together, while most other diagnostic systems use them only singly. Its process of diagnosis was to first isolate the problem to either the flight system or the ground and then further locate the offending subsystem. In parallel testing during Voyager activity, FAITH was reported to diagnose the Voyager ground system and spacecraft to a reasonable and useful depth (Friedman, 1983). FAITH was followed in 1987 by SHARP, an acronym for Spacecraft Health Automated Reasoning Prototype, which monitors telemetry, provides alarms if discrepancies are found between the anticipated and actual telemetry streams, and begins fault diagnosis activity. Current automated alarm systems monitor spacecraft telemetry, but they are static, simply alerting analysts when limits are violated. SHARP goes beyond the alert, comparing telemetry with expected activity based on the uplinked command sequence and providing a diagnosis of the cause of the discrepancy.

Meanwhile, at GSFC, advanced software development activities during the 1980s have resulted in the implementation of a real-time expert system into the operational environment of a spacecraft control centre. The program CLEAR, an acronym for Communications Link Expert Assistance Resource, is a fault isolation, rule-based expert system currently in use in the Payload Operations Control Center for the Cosmic Background Explorer (COBE) satellite. It is planned to also support the Gamma Ray Observatory after its launch in 1991. The COBE satellite uses the TDRSS for up to five 20-minute real-time communications contacts daily, which involve the uplink of stored commands, ranging, and receiving engineering telemetry. CLEAR monitors, in real time, more than 100 spacecraft and ground system performance parameters searching for configuration discrepancies and communications link problems. When it discovers an anomaly, CLEAR isolates it to as small a segment of the communications link as possible, notifies the analyst and provides advice toward effective resolution. As an advisory diagnostic expert system, CLEAR is a strictly passive component of the system, supporting (but not in-line with) spacecraft operations. Despite this passive nature, CLEAR has clearly demonstrated the utility and potential of expert systems in spacecraft mission operations (Hughes, 1989; Hughes, 1990).

One of the lessons learned from the development of CLEAR and its application to COBE is that while prototype expert systems may be developed rapidly, systems for actual operational use always require considerable, tedious development and testing, and may also require extensive modification for use by another spacecraft mission. To help solve this dilemma, GSFC has initiated development of an integrated, domain-specific tool, called GenSAA (Generic Spacecraft Analyst Assistant) that will allow spacecraft analysts to rapidly create simple expert systems themselves. GenSAA will use a highly graphical point-and-select method of system development to permit the analyst to use and modify previously developed rule bases and components to assemble the expert system for his application. This approach is one way to handle the increasing difficulty of mission operations that follows the growing complexity of spacecraft flight systems (Hughes, 1990).

Although it is very unlikely that the human being will ever be taken completely out of most functional processes, the ultimate goal of a super-automated system will be to limit human interaction to status requests and information displays, provided to give the person confidence in the automation devices, and to provide emergency override capability. This means processes would be initiated by other processes or functions; all interfaces, both internal and external, would be electronic, and no functions would require human intervention to complete. While we are a long way from this level of automation for an entire mission operations system, certain component functions may not be far off. Whatever the level of automation, it will be implemented incrementally, so that the proper testing can be completed and confidence developed prior to incorporation into actual operations.

8.5 Automated model-dependent data processing

The stringent data rate limitations placed on missions to other planets do not apply as strictly to Earth orbiters. With the advent of the TDRSS era, terrestrial satellites can downlink at rates of hundreds of megabits per second. Although the Hubble Space Telescope uses TDRSS and has a data rate which is high relative to past Earth-orbiting missions, the first mission to take real advantage of these high rates will be the Earth Observing System (EOS), a series of multi-payload platforms which will generate several gigabits of science data every day as they provide their investigators with the capability to study both land and sea surfaces as a function of time.

Generation of such large amounts of data requires a ground data system larger in scope than any previous to it. More important, EOS will attempt for the first time to automate parts of its data analysis through the use of invertible geophysical models. We have said in Chapter 5 that it is appropriate to archive the physical property being measured by a remote-sensing instrument — for example, spectral reflectance or distance. As currently conceived, EOS will depart from this practice by producing the geophysical observable under study as its standard archiveable data product. Spectral reflectance, for example, may be measured in order to determine mineralogic content of a surface. A measured spectrum may be compared to a catalogue of spectra to determine what minerals may be present in the sample, or a radar reflectance may be used to feed a numerical model of the surface which infers soil moisture, surface roughness or dielectric constant. Whereas in the past such models have been applied to project archives, EOS will adopt a set of agreed-to models and will archive the geophysical observable — composition, moisture, or roughness. Science data analysis is thereby made easier, as model-dependent processing can be supplied from within the mission, and a certain amount of data compression may be obtained as well. If the models are reversible, the primary data (i.e., the measured quantity) can be restored and need not be separately archived. If irreversible reduction is performed then either the primary data must also be archived or the mission must accept some loss in generality of its archive.

8.6 Standardization

Whether or not standardization of formats, processes and products can be called a future trend in the same light as the above five areas, it appears to be slowly happening due to necessity. Pressure to reduce the cost of operations has led to studies of the benefits to be gained if both the operational data and the data handling processes were designed from the beginning to conform to a set of standards. A study performed by the Mission Operations and Information Systems subcommittee of the Solar System Exploration Committee (Martin, 1986) concluded that there are seven areas where data system standards would contribute to cost savings. Several of them are well on the way to being implemented, but others have a lag time measured in years because they must be decided upon at the beginning of a project before the spacecraft and other data systems are designed. The seven are listed below.

(1) *Telemetry channel coding*. This standard would require all data coded on a spacecraft for telemetry downlink to use one of a predetermined set of coding algorithms.
(2) *Packetized telemetry*. This standard would require data generated by the science instruments and engineering subsystems on board spacecraft to conform to a common data structure, including frame size and format. Standards for packetized telemetry have been developed but are not in common use at this writing.
(3) *Packetized telecommands*. This standard would require all ground-prepared commands for transmission to a spacecraft to conform to a common data structure, including frame size and format.
(4) *Time code formats*. This standard would require all spacecraft and ground systems to use a common format for time, and to select that format from a predetermined set of formats. On-board clocks would be limited to specific oscillator frequencies, formats and characteristics.
(5) *Standard format data units*. This standard would require use of a common data structure for transfer of data between any and all elements of the ground data system. This can be retrofitted to an existing programme since it does not require changing in-flight formats, and in fact, both Magellan and Galileo have converted to a standard format data unit for file transfers between subsystems of the ground data system.
(6) *Telecommunications characteristics*. This standard would require usage of common frequency bands, ground timing stability criteria, and command, telemetry and ranging bandwidths between and within all facilities and agencies participating.
(7) *Tracking and orbit data formats*. This standard would require the use of a common data structure for spacecraft tracking data and a common set of conventions for the models and coordinate systems used to process the tracking data by all agencies participating.

Table 8.1 Key terms from Chapter 8

artificial intelligence	remote operations
distributed operations system	smart sensors
expert system	standard format data unit
heuristics	standardization
model-dependent data processing	teleanalysis
packetized telemetry	telecommanding
	telescience

Use of these standards would provide a mechanism for achieving effective cost reductions and greater efficiency in mission operations. Such standards would better enable the addition of automation for routine activities; permit more commonality in interfaces between programs; allow use of common data handling software; provide for a higher degree of flexibility to adapt to unforeseen circumstances during operations; allow for greater sharing of trained personnel and equipment between projects; and simplify the operational planning, implementation, testing and execution processes. With emphasis applied, perhaps these standards can be achieved during the decade of the 1990s.

8.7 Summary

This book has progressed from the basics of remote sensing, described in the first chapter, through the methodology for designing a mission operations ground system and the organization to be put in place to accomplish it, and through the details of the uplink and downlink processes. We have considered what to do when things go wrong and have concluded with a brief glimpse of future mission operations in this last chapter. The future appears replete with smart spaceborne vehicles, smart ground software in the form of expert systems, and small computers distributed all over the world in convenient experimenters' labs and engineers' control rooms where control of the instrument or subsystem is a finger's touch away. Some of these concepts have been introduced briefly in this chapter; key terms are listed in Table 8.1. It is our hope that the excursion through this book has provided the reader with some of the basic understanding and tools to help make this future vision for remote-sensing mission operations systems both a technical and an economic reality.

8.8 Exercises

(1) For an Earth resources satellite in a 10-hr polar orbit, whose platform control centre is at GSFC, what are some advantages and disadvantages of

having the control centre for the prime instrument of the three constituting the payload located at the manufacturer's facility in Europe? Can you devise some innovative ways to circumvent some of the disadvantages?

(2) For the mission in Exercise 1, describe how resolution of conflicts between instruments over payload resources might be accomplished if the prime instrument is controlled from Europe.

(3) For each of the 21 level 2 functions in the functional decomposition of Figure 2.5, state whether distribution to a remote location would be desirable for a typical multi-instrument, Earth-orbiting science satellite mission, or whether the function should remain centralized. If you believe there is not enough information to make this decision, specify the nature of the additional information needed to allow such a decision to be made.

(4) Construct an information flowchart, similar to Figure 7.5, for a distributed MOS in which requests are input through computers in remote universities, activity plans are negotiated between an instrument control centre and the remote experimenters until conflicts are resolved, and final activity plans are sent back to the university for approval. Add to the drawing the interfaces necessary for university personnel to browse through a central library of metadata, select an image for analysis, and transfer the full-resolution image to the local computer.

(5) Between 1982 and 1991, the typical personal computer purchased for home or small business application grew from a 64 kilobyte, 8-bit processor unit with dual floppy disk drives to a unit with several megabytes of memory, a 16- or 32-bit processor and a 128 megabyte hard disk drive. Assuming it is only a matter of time until megabytes of storage becomes flight qualified, discuss some of the benefits and some of the pitfalls that await the mission operator of the future with megabytes of memory at their disposal.

(6) A future weather satellite is designed with on-board smart sensors to detect severe atmospheric weather and increase the data sampling frequency tenfold. What complications does this create for the mission operations system? Assume that the satellite is in an orbit that precludes continuous contact with a ground station, and therefore that onboard mass storage is provided.

(7) Of the six expert system programs used for spacecraft operations mentioned in this chapter, only one, CLEAR, has been implemented into a mission operations environment, and it is strictly passive. No error that the program might make can result in a error on-board the flight vehicle. What types of assurance (testing, analyses, history, parallel operations, etc.) do you believe the mission controllers must have to permit an expert system to autonomously send commands to a vehicle?

(8) Which of the seven items recommended by the Solar System Exploration Committee for standardization do you believe would have the highest pay-

back in terms of cost savings if universally implemented? You may choose more than one, but justify your selections.

References

Eggemeyer, W. C. and Cruz, J. W., 1990, 'PLAN-IT-2: The next generation planning and scheduling tool', presentation at Goddard Conference on Space Applications of Artificial Intelligence, Baltimore MD, May.

Friedman, L., 1984, Diagnosis Combining Empirical and Design Knowledge, presentation at the American Association for Artificial Intelligence, Austin TX, August.

Hughes, P. M., 1989, 'CLEAR: Automating control centers with expert system technology', presentation at the NASA Johnson Space Center Space Operations, Automation and Research (SOAR) Conference, Houston TX, July.

Hughes, P. M., 1990, 'Utilizing expert systems for satellite monitoring and control', presentation at the Space Operations, Automation and Research (SOAR) Conference, Albuquerque, NM, July.

Hyman, R., 1984, *Command and Control Operations Concept Study, Levels of Automation*; Computer Technology Associates, Inc., GSFC Contract NAS5-27684, Task 500-03c (July 1984), Greenbelt, MD: National Aeronautics and Space Administration.

Ledbetter, K. and Taliaferro, D., 1983, *Conceptual Design for a Transportable Distributed Command and Control (TDCC) System*, Design Concept Document, Computer Technology Associates, Inc., GSFC Contract NAS5-27300 (Oct 1983) Greenbelt, MD: National Aeronautics and Space Administration.

Martin, J. S., (Chairman), 1986, *Planetary Exploration through Year 2000, A Core Program: Mission Operations*, A Report by the Solar System Exploration Committee of the NASA Advisory Council, Mission Operations and Information Systems Subcommittee, Washington DC: National Aeronautics and Space Administration.

Rokey, M. and Grenander, S.,1990, Planning for space telerobotics: the remote-mission specialist, *IEEE Expert, Intelligent Systems and their Applications*, **5**, 3.

Vere, S., 1983, Planning in time: windows and durations for activities and goals, D-527, IEEE Transactions on Pattern Analysis and Machine Intelligence **5**, 3.

Wallis, B. D., 1982, *Functional Requirements of Semiautomated Sequence Integration Software*, 715–154, internal document Pasadena CA: Jet Propulsion Laboratory.

Appendix 1.

The Computer Software Management and Information Centre

Computer software development can be a significant expense in the entire development of a remote-sensing mission. So that duplication of such efforts might be reduced, some of the programs used in NASA-supported missions, including the SMDOS, SINDA and TRASYS programs discussed in this book, are available through the Computer Software Management and Information Centre (COSMIC). COSMIC is operated for NASA by the University of Georgia for the purpose of making programs available to the public at nominal cost. COSMIC catalogues may be obtained at the following address:

COSMIC
382 East Broad Street
University of Georgia
Athens, GA 30602
USA

Programs we have referenced in this book are catalogued by COSMIC as follows:

SMDO	SNPO-16933
SINDA-85	MSC-21528
TRASYS-II	MSC-21030

Appendix 2.

Acronyms and Abbreviations

AEDR (Magellan) Archive Engineering Data Record
AFRTS Air Force Remote Tracking Station
AODR Analog Original Data Record
ASPP (Magellan) Advanced Sequence Planning Package
ATR Acceptance Test Review
BOT Beginning Of Track (on a tape recorder, or related to a tracking antenna's active period)
C-MIDR (Magellan) Compressed Mosaicked Image Data Record
CDR Critical Design Review
CMD Command
COBE Cosmic Background Explorer
DMAT (Magellan) Data Management and Archive Team
DN Data Number
DODR Digital Original Data Record
DOY Day Of Year (numbered from one to 365)
DSN Deep Space Network
DSS Deep Space Station
EDR Experiment Data Record
EU Engineering Units (see Glossary)
EOS Earth Observing System
F-BIDR (Magellan) Full-resolution Basic Image Data Record
FEM (Magellan) Forward Equipment Module
FRD Functional Requirements Document
GMT Greenwich Mean Time
GPS Global Positioning System
GSFC NASA Goddard Space Flight Center
HGA High Gain Antenna
IDPT (Magellan) Image Data Processing Team
ISOE Integrated Sequence of Events
JPL NASA Jet Propulsion Laboratory
MCT (Magellan) Mission Control Team
MD Mission Director
MGA Medium Gain Antenna
MOM Mission Operations Manager
MOS Mission Operations System
MSDT (Magellan) Mission and Sequence Design Team

NASA National Aeronautics and Space Administration
NAV Navigation
ODR Original Data Record
OPCT (Magellan) Operations Planning and Control Team
ORR Operational Readiness Review
OWLT One-Way Light Time
PCR Preliminary Concept Review
PDR Preliminary Design Review
PIO Public Information Office
PM Project Manager
POCC Payloads Operation Control Centre
PRR Preliminary Requirements Review
PSG Project Science Group
RSET (Magellan) Radar System Engineering Team
RTG Radioisotope Thermoelectric Generator
RTLT Round-Trip Light Time
S/A Solar Array
S/C Spacecraft
SAMPO (Magellan) Science and Mission Planning Office
SAR Synthetic Aperture Radar
SCT (Magellan) Spacecraft Team
SDPT (Magellan) SAR Data Processing Team
SEDR Supplementary Experiment Data Record
SEF Spacecraft Events File
SEU Single Event Upset
SFOS Spaceflight Operations Schedule
SIR Shuttle Imaging Radar (SIR-A, SIR-B, SIR-C)
SPFPAD (Viking) Spacecraft Performance and Flight Path Analysis Directorate
SPICE (see Glossary)
SRR Systems Requirements Review
STDN Satellite Tracking and Data Network
SVL (Magellan) Systems Verification Laboratory
T-BIDR (Magellan) Temporary F-BIDR
TDR Team Data Record
TDRSS Tracking and Data Relay Satellite System
TEDR (Magellan) Temporary Experiment Data Record
TLM Telemetry
TT and C Telemetry, Tracking and Command
TWTA Travelling Wave Tube Amplifier
USAF United States Air Force
UTC Universal Time, Coordinated
VOIR Venus Orbiting Imaging Radar
X-BIDR (Magellan) Expedited F BIDR

Appendix 3.

Glossary

A-, B-, and C-specifications, terminology used for various levels of requirements specification documents by US military mission contracts. The A-specification is the top-level requirements document for a system. B-specifications are for major subsystems, and C-specs are written for each individual component.

activity, a collection of one or more actions to perform in order to accomplish a goal or objective. It may relate to actions requested of a remote-sensing payload or to ground actions in support of a remote-sensing mission operations system.

activity plan, a time-ordered listing of activities which can be and are intended to be accomplished by a mission or a phase of a mission.

adaptive, a term to describe a mission capable of responding in a timely way to discoveries made in its own remote-sensing data.

advanced sequence plan, See *activity plan.*

alarm, *n.* a condition which occurs when the value of a downlink telemetry parameter exceeds a pre-defined limit; (*v.*) to monitor a telemetered parameter for some specified condition and to identify it (typically by a visible or audible signal) when it violates (or achieves) that condition.

alarm limit checking, a function performed on real-time telemetry to verify that the specific value of a given measurement is within pre-defined bounds.

alternate mission, a plan which describes an altered set of mission goals following the occurrence of an anomaly and the execution of a contingency plan.

altimeter, an instrument whose function is to measure the altitude of a platform above a reference surface.

ancillary data, data valuable for analysis which are to be associated with user data.

anomaly, the actual or potential occurrence of an unexpected event having an impact on subsequent operations. The word is used to describe the root cause of the occurrence rather than the direct observable. See *symptom.*

apoapsis, that point in a satellite's orbit which is most distant from the primary's centre.

archive, a repository designed for the permanent storage and distribution of data and/or metadata.

archive copy, a copy of a data product designed for a deep archive only and not for access except in special situations. See *deep archive*.

artificial intelligence, a field of computer science concerned with the concepts and methods of programming computers to make decisions based on information, logic and deductions that humans recognize as properties of intelligence.

attitude, the angular orientation of a platform, such as a spacecraft, in flight.

attitude control subsystem, that portion of a platform whose function is to control the orientation of the platform in space.

B-specification, See *A-specification*.

bit error rate, the average number of bits per bit group in a digital data transmission stream that are corrupted; the specification on how many are permitted to be corrupted.

block, an ordered group of commands that, when executed, will perform a specific activity.

C-specification, See *A-specification*.

caution and warning, See *alarm*.

clock kicks, a technique in simulation where the simulated clock is advanced past periods of low activity to minimize verification time.

command database, a spacecraft-unique data file resident in the ground command system which provides the correct binary bit pattern for each valid command and command option.

command interpreter, a form of simulation that interprets each command and displays the expected result.

command simulator, a software or hardware model of a platform or instrument which, when a command file is input, will indicate how the actual entity will respond to given commands.

commutation, the platform process of assembling ordered sets of repeating telemetry values, with varying number of bits each, into a telemetry frame for downlink.

commutation map, a scheme describing how data are to be commutated.

compression, the process of minimizing the number of bits used to transfer information. Compression may be either lossless, where no information is sacrificed, or lossy, where information is intentionally lost.

concept review, a formal review, before a board of experts, of an operations concept for a mission.

configuration control, a rigorous process established for a project to control changes to design, implementation, build, test and operations, and their associated documentation and databases.

contingency plan, a plan describing action to be taken should the subject anomaly occur.

contingency response, a response or potential response to an anomaly.

cumulative constraint, a constraint whose limitation is applied on a cumulative bases, such as fuel usage.

cyclic, a repetitive sequence of command blocks representing an activity that is called many times.

data, a collection of information; plural of datum, a unit of information.

data flow diagram, a diagram, drawn to specific standards, that identifies the data interfaces between the functions derived from a functional analysis.

data numbers (DN), numbers taken directly from digital telemetry without conversion to *engineering units*. Data numbers are unitless and generally range from zero to some even binary multiple less one (2^n-1).

datatake, a single data-gathering event to be scheduled; for example, in the case of an camera, a single image.

decalibrate, to convert telemetry, generally in the form of integer data numbers (DN), into engineering units (EU) representing geophysical units such as voltage, angle, reflectance, etc., which will have the range and granularity defined by calibration curves or tables. See both *data numbers* and *engineering units*.

decommutate, to disassemble a downlinked telemetry frame into its constituent measurements, each of which may use a different number of bits to represent the value.

Deep Space Network, a network of tracking antennas and support stations built and operated by NASA for the tracking of planetary spacecraft. Deep Space Stations are located near Madrid, Spain, Canberra, Australia, and Barstow, California, USA (See Figure 1.10).

deep archive, an archive which is designed for long-term storage of valuable data and not used for access to that data except when the working copy of the data is damaged or lost.

deep archive copy, See *archive copy*.

design review, any of several formal reviews, before a board of experts, where the design of an element (or the entire system) is critiqued prior to allowing implementation of the design to begin.

design specification, written requirements for what the design must accomplish.

discrete constraint, a constraint whose limitation is applied on an instance-by-instance basis, such as usage of a single component.

distributed system, an operational system where one or more functions are physically removed to a location remote from other functions such that all voice and data interactions must be accomplished indirectly through an electronic medium.

document tree, a structured diagram of all documents (or document categories) in a project for the purpose of illustrating dependencies of one on another.

Doppler shift, a shift in the frequency of electromagnetic radiation due to relative motion of transmitter and receiver.

downlink, the data flow which begins with an activity aboard a platform and ends with receipt of the data by the intended user; sometimes used to refer to only the telecommunications transmission portion of this data flow.

downlink station, the facility for the antenna and its co-located equipment used to receive signals from a platform.

electrical power subsystem, the platform subsystem providing power to other subsystems, payloads and instruments.

encoding, a scheme of representation of data in a digital stream that protects against bit errors.

engineering data, data collected for spacecraft engineering purposes, to monitor and analyse the condition of the platform or payload. See also *science data*, *user data*.

engineering units (EU, often eu), units of measure which have direct engineering or scientific meaning, such as volts, metres, watts, etc.

ephemeris (*pl.*: **ephemerides**), a time-ordered set of data describing the location or position of an object in space (e.g., a spacecraft).

event, an occurrence of an activity.

experiment data record, a data product consisting of science data which have been separated from the telemetry stream and re-organized into a form more useful to the users.

Experiment Representative, a position on a mission operations flight team serving as an interface between the scientists and flight team engineers.

expert system, a computer program that performs a specific task by operating on a large knowledge base using symbolic inference and logical rules to arrive at a statistically probable answer to a query or problem.

fault protection, software responses of a flight subsystem to its own perception of a fault having occurred on board with the intent of solving the fault or safing the flight system.

fault tolerance, the ability of a system to operate under a given number of failures or anomalies. A two-fault tolerant system is required to be capable of operating in the presence of two simultaneous, independent faults.

flight controllers, a generic term referring to mission operations system personnel who perform real-time platform support functions, in particular, the monitoring of telemetry displays and the sending of commands.

flywheeling, the capability of a frame synchronization (*q.v.*) function to utilize knowledge of frame lengths to jump ahead in its search for the synch word to just prior to its estimated position in the telemetry stream.

frame, a group of bits in a telemetry stream, delineated by an initial fixed bit pattern and representing a repeating set of measurements. See *synchronization word*.

frame synchronization, the processing of real-time telemetry to locate and mark the beginning and ending of each group of bits starting with a fixed-pattern frame synch word.

free-flyer, a spacecraft whose primary purpose is to serve a limited number of payloads.

function, operational processes performed by hardware, software or people; a collection of activities.

functional analysis, a study of a system through the examination of its elements as described by the system's functions.

functional decomposition, a method of performing functional analysis by breaking down top-level functions into their constituents, and breaking these down further into more detailed elemental functions.

functional hierarchy, a diagrammatic technique of presenting a functional decomposition by showing in layered levels the functions of a system.

functional requirements, requirements that govern the design of the system components to satisfy operational requirements by identifying the specifications for each function the component must accomplish.

ground, a euphemism referring to components, personnel or operations supporting an in-flight platform that occur on the Earth.

handover, the period during which two entities share an activity as one ends the activity and the other begins it, as when one tracking antenna ceases to receive or transmit signals while another begins, or when one personnel shift ends and another begins.

heuristics, the process of developing rules learned through discovery.

hysteresis, the suppressing of an alarm on a telemetry measurement until the alarm condition is met by more than a specified number of consecutive samples.

input, information necessary to begin a process or function.

intensity information, information about a target relating to the amount of electromagnetic radiation it returns.

interface, information flow between two functional components of a system.

interface analysis, a process by which information flow between functions of a system is analysed.

interface requirements, a specification of the characteristics of information flow between functions of a system.

inverse planner, a software tool to identify all possible datataking opportunities.

light time, the time it takes for electromagnetic energy (e.g., light or radio waves) to travel the distance from a platform to Earth.

logical functional model, a model of a system where the division into functions is based on required activities only, avoiding physical orientations such as specific hardware, software, facilities or existing institutional elements.

lossless compression, See *compression.*

lossy compression, See *compression.*

macro, a grouped set of commands or other instructions to perform a specific activity.

malfunction procedure, See *contingency plan.*

memory management, refers to the locations in on-board memory in which flight software, parameters and command sequences are stored and the time order of moving the contents of memory around to perform necessary operations.

metadata, information about data, such as a catalogue of data products.

mission, a major task assigned by a sponsor; the largest division of work considered in this book.

mission control, the functional element of a mission operations organization responsible for real-time monitoring of the platform and day-to-day operations of the ground control centre. See also *flight controllers*.

mission management, the functional element of a mission organization responsible for typical management functions.

mission operations, the collection of all activities on the Earth that support the objectives of a specific remote-sensing in-flight platform and payload.

mission operations system, the collection of hardware, software, people and documentation required to perform, monitor and control the operation of a mission.

mission planning, the first (in time) task in the uplink process, where activities requested by users are defined in terms of required resources and analysed to see if they can be implemented.

mission planning and scheduling, the functional element of a mission responsible for planning on a longer scale than real time.

mission requirements, high-level statements of the goals and objectives of a remote-sensing mission that specify what it is the mission is required to achieve.

***N*-squared chart**, a diagrammatic technique to display the results of functional interface analysis.

nadir, toward the surface of a planet which is directly under an orbiting satellite, or referring to the surface point directly under it.

one-way lock, the synchronization of the frequency of signals transmitted by a platform to the frequency at which the ground receiver is set so that telemetry might be received.

operability, a measure of the ability of a system to be operated.

operational requirements, requirements on a system relating to the methods of achieving the operational mission goal.

operations concept, a document containing an orderly collection of user-oriented ideas as to how the mission operations system should function to satisfy the mission and experiment objectives.

operations plan, a document describing the intended team organization and structure, roles and responsibilities, and high-level activity descriptions of all the activities to be performed by the organizational element covered by the plan.

operations procedures, a set of detailed step-by-step lists of tasks defining how each operational activity is to be accomplished, often containing check-lists to indicate task completion as they are finished.

operations readiness review, a formal review, before a board of experts, of the readiness of a flight team to begin mission operations.

opportunity timeline, a software tool to identify or define datataking opportunities.

original data record, the first recording of downlink telemetry. Sometimes separated into analog original data record, an analog recording of the antenna signal, and digital original data record, a recording of the digitized data.

output, information resulting from performing a process or function.

packet, a constant-length group of bits used in standardized telemetry systems.

packetized, as applied to telemetry, divided into individual units (packets) of the same frame size and design.

parity, specification of oddness or evenness of the sum of all the ones in a group of bits (zeroes are not counted). An odd number of ones means that the bit group has odd parity; an even number of ones means that the bit group has even parity.

particle detector, a payload instrument which measures properties of small particles in space or a planetary atmosphere.

payload, a remote-sensing instrument or collection of related instruments flown aboard a platform for the purpose of obtaining scientific data.

performance requirements, requirements specifying when, or how often, functions or activities must be completed and their duration.

periapsis, that point in a satellite's orbit which is least distant from the primary's centre.

phase, one of the major time divisions of a mission, e.g., launch, encounter.

physical functional model, a model of a system where the division into functions takes into account physical orientations such as specific hardware, software, facilities or existing institutional elements.

pixel, the individual picture elements of a picture, each of which is digitized for transmission and reassembled during science data processing.

platform, that element of a flight system which carries and supports payloads.

process, an ordered collection of activities that accomplish one or more functions in a mission operations system.

propulsion subsystem, the platform subsystem which can move the platform in both attitude and position by expelling gasses through a nozzle.

radio science, a science which determines properties of planetary atmospheres or other phenomena by recording the strength and frequency of the platform's telecommunications signal after its path passes through or near the object of investigation.

radio subsystem, See *telecommunications subsystem.*

radiometer, a device for measuring the energy content of electromagnetic radiation.

rate constraint, a constraint whose limitation is applied on a rate basis, such as power usage.

real time, in the immediate time period, or directly associated with the origin, as in real-time data.

remote sensing, the act of acquiring scientific data concerning a target without contacting it.

requirements review, one of several formal reviews, before a board of experts, to assess whether or not the system's requirements definition is sufficient to allow the design to begin.

resource, any quantity utilized in order to accomplish an activity.

safing, an action taken to return a platform or payload to a safe condition.

scan platform, a rotatable structure which is used to aim instruments at their targets.

scatterometer, a device for measuring the scattering properties of targets.

scenario, a 'what-if' example of usage of a mission operations system, where details of a specific mission are postulated using the concept, the design and the plan, to see if any flaws in the design logic can be uncovered.

Science Coordinator, See *Experiment Representative.*

science data, data acquired for the science community, generally the data taken by the payload instruments. See *user data.*

script, See *sequence of events.*

sensor data processing and analysis, the functional element of a mission that processes downlinked data into products and analyses them.

sequence, an ordered set of spacecraft commands and/or command blocks.

sequence design, the second major task in the uplink process, in which groups of commands are organized in order to accomplish each entry in an activity plan.

sequence of events (SOE), a time ordered listing of merged inflight and ground events used by flight teams to monitor operations activities.

sequence translation, the third major task in the uplink process, in which sequences of commands are translated into bit-level instructions relevant to a particular element of the flight system.

simulator, See *command simulator.*

single event upset, an occurrence caused by cosmic rays in which a bit within a flight computer memory or register is temporarily or permanently altered.

sink, the destination device or position of a data stream. See *source.*

slice, a removable unit in an electronic flight component.

source, the originating device or position of a data stream. See *sink.*

spacecraft health and safety, the functional element of a mission responsible for monitoring, analysis and maintenance of flight hardware and software.

spatial information, information about a target relating to the physical relationship of its features.

spectral information, information about a target relating to differences in the ways in which it scatters different wavelengths of electromagnetic radiation.

spectrometer, a device for measuring spectral information.

SPICE, an acronym representing the planetary archiving concept. Those areas where archiving is required are *S*atellite ephemeris, *P*robe ephemeris, *I*nstrument data, *C*ommand data, and *E*vent data.

state model, a simulator which mimics the state of every part of platform and payload as a function of time.

subcommutation, a technique used in telemetry where different measurements occupy the same place in a telemetry frame or packet on a cyclic basis.

supplementary experiment data record, a separate data record of information related to datatakes, such as ephemerides, pointing parameters, etc., which assists the experimenter in interpreting his instrument's data.

symptom, a directly observable non-nominal situation, without regard to its root cause. See *anomaly.*

synchronization word, pseudo-random bit pattern at the beginning of each telemetry data frame, used for frame synchronization.

synthetic aperture radar, a remote-sensing technique using microwave pulses to form an image.

system, a major mission component.

system development, the designing and building of a system; the second major phase of a system life cycle.

system integration, the melding together of system components into a functioning whole; part of the third major phase of a system life cycle.

system operations, the actions associated with making a system work to accomplish its intended goal; the fourth major phase of a system life cycle.

system planning, the conceptual and feasibility analysis to define what a system must accomplish; the first major phase of a system life cycle.

system test, the verification of functionality of a system; part of the third major phase of a system life cycle.

target, an entity which is to be studied, or from which remote-sensing data are to be taken.

team data record, a product containing science data which have been processed into some specialized form, such as an image which has been specially processed.

telecommunications subsystem, the platform subsystem responsible for communication between platform and ground.

telemetry, data which are transmitted via radio link; usually applied only to digital data.

thermal subsystem, that element of a platform responsible for controlling the on-board thermal environment.

timeline, a logical, time-ordered sequence of activities.

travelling wave tube amplifier, a device which creates high-power radio signals at microwave frequencies.

two-way lock, the establishment of a coordinated command and telemetry link between an in-flight platform and a ground antenna involving synchronization of transmitter and receiver frequencies.

uplink, the data flow which begins with a user's intention for a platform activity and ends with the accomplishment of that activity; sometimes used to refer to only the telecommunications transmission portion of this data flow.

uplink station, the ground antenna being used to transmit signals to a platform.

user, the individual or other entity for whom remote-sensing data are taken.

user data, data acquired specifically for the user. See *science data*.

validation, the process of verifying that a command, block or sequence will execute the intended actions after it is received on board.

variable packet, a telemetry packet in which different measurements are subcommutated at different times.

very long baseline interferometry, a technique to precisely determine the location of a spacecraft at planetary distances by simultaneous tracking with two or more ground antennas well separated on the Earth's surface.

voice network, a system of telephone handsets, headsets, speakers and communications lines to allow operational stations to talk instantly to each other without the necessity of first obtaining access to a line.

wheel desaturation, the act of slowing momentum wheels while holding a spacecraft's attitude constant.

working copy, a copy of a data product designed for access.

Index